"This is an interesting, non-mathematical treatment of the subject. Prof. McGlinchey has an excellent reputation as a researcher and teacher internationally."

J J Vittal
Emeritus Professor, National University of Singapore, Singapore

"This book illustrates how chemical reactivity is controlled by molecular and electronic symmetry. Professor McGlinchey applies symmetry-based arguments to explore the origins of well-known phenomena which illustrate fundamental principles, and then use these examples to thread together the arcana of organic, inorganic and organometallic chemistry. What makes this approach so appealing and effective is that it brings together seemingly disparate subjects, crosses traditional discipline-based boundaries and allows the building of conceptual bridges. Complex arrays of structural patterns and chemical behavior are linked by simple unifying ideas, which lighten the otherwise overwhelming burden of facts that fill most chemistry textbooks. In addition, McGlinchey 'humanizes' his analysis by providing vignettes of the lives of some of the individuals whose discoveries molded the way we think about chemistry. The refreshingly light and catholic nature of its presentation will be an exciting and useful read for all those interested in the way the molecular and electronic structure control chemistry."

Richard Oakley
Emeritus Professor, University of Waterloo, Canada

Making and Breaking Symmetry in Chemistry

Syntheses,
Mechanisms
and Molecular
Rearrangements

Making and Breaking Symmetry in Chemistry

Syntheses, Mechanisms and Molecular Rearrangements

Michael J McGlinchey

University College Dublin, Ireland

World Scientific

NEW JERSEY · LONDON · SINGAPORE · BEIJING · SHANGHAI · HONG KONG · TAIPEI · CHENNAI · TOKYO

Published by

World Scientific Publishing Co. Pte. Ltd.

5 Toh Tuck Link, Singapore 596224

USA office: 27 Warren Street, Suite 401-402, Hackensack, NJ 07601

UK office: 57 Shelton Street, Covent Garden, London WC2H 9HE

Library of Congress Cataloging-in-Publication Data

Names: McGlinchey, Michael J., author.

Title: Making and breaking symmetry in chemistry : syntheses, mechanisms and
    molecular rearrangements / Michael J. McGlinchey, University College Dublin, Ireland.

Description: New Jersey : World Scientific, [2022] | Includes bibliographical references and index.

Identifiers: LCCN 2021054615 | ISBN 9789811249655 (hardcover) |
    ISBN 9789811249662 (ebook for institutions) | ISBN 9789811249679 (ebook for individuals)

Subjects: LCSH: Molecular theory. | Symmetry (Physics)

Classification: LCC QD461 .M397 2022 | DDC 541/.22--dc23/eng20220228

LC record available at https://lccn.loc.gov/2021054615

British Library Cataloguing-in-Publication Data

A catalogue record for this book is available from the British Library.

For any available supplementary material, please visit
https://www.worldscientific.com/worldscibooks/10.1142/12652#t=suppl

Desk Editor: Shaun Tan Yi Jie

Typeset by Stallion Press
Email: enquiries@stallionpress.com

Preface

Symmetry, in particular broken symmetry, underpins much of our current understanding of natural phenomena. Focussing here on chemistry, our objectives are two-fold: first, to demonstrate the underlying significance of symmetry arguments in the experimental elucidation of the mechanisms of chemical reactions and rearrangements in the organic, organometallic and inorganic domains; second, to show that while the syntheses of a number of molecules of unusually high symmetry have been spectacularly achieved, others continue to pose serious challenges.

The methods of determining reaction mechanisms have evolved greatly since the approaches developed by the early pioneers. Many now-classic experiments carried out more than three-quarters of a century ago took advantage of the retention, inversion or loss of chirality, or the incorporation of a radioisotope then available, such as carbon-14 or iodine-128. These included the demonstration of inversion of configuration in S_N2 reactions, the existence of intermediate species such as benzynes, cyclopropanones, carbenes, oxirenes, halogen-bridged carbocations, nitrilium ions, and also the discovery of a number of synthetically important molecular rearrangements (pinacol, Favorskii, Wolff, Beckmann, Curtius, benzidine). Those researchers relied on their outstanding laboratory technical skills to develop approaches that have since been adopted, modified and improved by using currently available technology.

After reminding ourselves of a number of these famous pioneering experiments, we then discuss more recent advances made possible by the availability of modern instrumentation, such as infrared and nuclear magnetic resonance (NMR) spectrometers with their enormously enhanced sensitivity and multidimensional capabilities. Typically, the use of ^{14}C isotopic labels with the requirement of radiochemical detection techniques has been largely superseded by ^{13}C NMR studies, whereby the progress of reactions can be conveniently followed continuously over a wide range of temperatures. Among the particular cases in which judicious symmetry breaking played a role, we highlight the elucidation of the mechanism of alkene metathesis, the subtleties of carbonylations or nucleophilic substitutions in organometallic systems, the 2-norbornyl cation controversy, the use of chiral methyl ($C^1H^2H^3H$) or chiral phosphate ($P^{16}O^{17}O^{18}O$) tripods in biological systems, electron transfer processes in transition metal complexes, and molecular rearrangements in sulfur-nitrogen clusters and carboranes.

Another section is devoted to the detection and elucidation of otherwise hidden rearrangement processes, or reactions in which the detailed mechanism was obscure. In these cases, carefully designed symmetry breaking was utilised whereby any perturbation of the molecular geometry or reactivity was minimised. This was frequently accomplished by the incorporation of suitable NMR probes, by which means the process could be monitored continuously over a wide range of temperatures, thus allowing the extraction of activation energy data.

While the seminal contributions of Woodward and Hoffmann explicating the principles of the Conservation of Orbital Symmetry provided a fundamental understanding of many aspects of chemical reactivity, they also made numerous predictions, and experimental verification of these concepts has, in many cases, been achieved by the astute, sometimes rather subtle, breaking of molecular symmetry. One should never forget that the real test of a theory is whether it matches the experimental result. Moreover, one should not undervalue the contributions of the synthetic chemist in this regard; it is still easier to envisage a particular molecule, suitably labelled for our purposes, than it is to prepare and characterise an analytically pure sample of it. Chemistry is fundamentally still an experimental science!

The continual advances in computer technology and computational methods have transformed our approaches to many problems; in particular, their significance in X-ray crystallography has been pivotal. The development of more powerful X-ray sources, in conjunction with enhanced data collection systems and the newer direct methods programs, have reduced acquisition times and structural determinations from many days to just a few hours, thus allowing the ready characterisation of an entire reaction sequence.

Inevitably, the choice of topics may be influenced to some extent by the interests of the author, but an attempt has been made to take examples from all across the Periodic Table, selected from the literature published in recent decades. These are intended to show clearly how symmetry breaking in conjunction with the currently available spectroscopic, spectrometric and other analytical techniques frequently provide an approach whereby mechanistic proposals for reaction mechanisms and molecular rearrangements can be more carefully evaluated.

The latter part of the book is devoted to discussion of the synthetic routes to highly symmetrical molecules, including those that parallel the structures of the Platonic solids, the tetrahedron, octahedron, cube, icosahedron and dodecahedron. Moreover, we also describe the attempts (some, as yet, only partially successful) to prepare the hydrocarbons, C_nH_n, where n is 4, 6, 8, ..., 20, that are the inverse polyhedra of the known closo borane anions, $[B_xH_x]^{2-}$, where x is 5 through 12. While some organic molecules exhibiting high or unusual symmetry are now accessible in kilogram quantities (cubanes, corannulene), and C_{60} with its soccer ball symmetry is now available on a commercial scale merely by the vaporisation of graphite, others such as tri- or penta-prismane are still niche products, available only in small amounts, and with considerable difficulty. In contrast, in the inorganic and organometallic domains where smaller bond angles are more easily tolerated, systems of tetrahedral, prismatic or cubic symmetry are numerous.

The book is deliberately written in a relatively informal, almost conversational, style aimed at providing senior undergraduates and graduate students with both a historical and a more up-to-date survey of the elucidation of many important reaction mechanisms and molecular rearrangements; it also recognises and celebrates the contributions of the early pioneers. In addition, one hopes that it may prompt some of the more established researchers to adopt, modify and improve the approaches described here, as well as attract their attention towards some currently unavailable synthetic targets. Many of the publications cited are discussed only briefly, and from the particular perspective of the symmetry properties of the molecules therein; however, since they contain much other valuable information for the reader who seeks more detail and depth, a comprehensive list of literature references is provided.

About the Author

Dr. Michael J. McGlinchey is Emeritus Professor at University College Dublin, Ireland where, after spending 30 years at McMaster University in Canada, he became Chair of Inorganic Chemistry (2002–2010), Head of Department (2003–2005), Head of School (2005–2007), and Director of Dublin Chemistry (2008–2010). He is an elected Fellow of the Chemical Institute of Canada (since 1985) and Royal Irish Academy (since 2008). He is a recipient of the Alcan Award in 2000 from the Canadian Society for Chemistry for distinguished contributions to inorganic chemistry. He has published about 300 papers in the areas of organic, organometallic and bio-organometallic synthesis and mechanisms, sterically hindered molecules, NMR fluxionality, and X-ray crystallography. He has held visiting professorships in France (Rennes, Paris, Versailles, Toulouse), Switzerland (Geneva, Lausanne), People's Republic of China (Heilongjiang, Siping) and the National University of Singapore. He holds a PhD from the University of Manchester, UK.

Contents

Chapter 1

Introduction

"The chief forms of beauty are order and symmetry."

— *Aristotle*

Symmetry can be viewed from the artistic or mathematical perspectives. We will here take an intermediate position, acknowledging the creativity of chemical synthesis while taking advantage of the formalism provided by those applications of Group Theory that are most relevant to chemistry. An obvious example of symmetry breaking is the phenomenon of chirality, so evident in our everyday life in gloves or corkscrews, in Nature in conch shells (Figure 1.1), at the molecular level in left-handed amino acids or right-handed DNA helices, or at the subatomic domain in left-handed neutrinos. In everyday objects or in molecules, chirality is characterised by the non-superimposability of their mirror images. However, we can also categorise molecules more formally in terms of their symmetry elements, and how these are modified in the course of a chemical reaction, or during internal rearrangement.

Of course, chiral systems are ubiquitous in natural products such as alkaloids, steroids, isoprenoids, peptides, etc., as well as in molecules containing moieties such as (R)-(+)-2,2′-bis(diphenylphosphino)-1,1′-binaphthyl (BINAP),[1] specifically designed for use in asymmetric catalysis. Such systems have been reviewed frequently and

1

Figure 1.1. A left-handed glove, a right-handed corkscrew and a pair of conch shells.

comprehensively by authorities in those particular fields, and are not our focus herein. Instead, we aim to discuss the syntheses, rearrangements and dynamic behaviour of simple organic, inorganic and organometallic molecules in which the investigation and control of symmetry plays a crucial and specific role. As part of this endeavour, we also include occasional brief vignettes of historical interest as well as those in relevant topics in areas such as stereochemistry or NMR spectroscopy; these are clearly marked in boxes.

When discussing reaction mechanisms, molecular rearrangements and syntheses of highly symmetrical molecules, we shall routinely classify such processes in terms of their symmetry elements and point groups. For those unfamiliar with, or who may have forgotten, these concepts we provide a brief summary in an Appendix. However, for more in-depth understanding, consultation of one or more of the many excellent books expounding the applications of symmetry and group theory to chemistry is highly recommended. In this introductory section, we make some general comments about reaction mechanisms, syntheses and experimental techniques, and briefly indicate those areas that will be further explored in the remaining chapters. We also briefly highlight the lives and achievements of some of the early pioneers in these endeavours.

1.1. Symmetry Breaking — Werner's Remarkable Insight

The incorporation of different chemical elements into homogeneously substituted molecules inevitably brings about a lowering of their symmetry. Thus, successive replacement of a single chlorine in CCl_4 (T_d) by hydrogen to generate chloroform, $CHCl_3$ (C_{3v}), and again to form dichloromethane, CH_2Cl_2 (C_{2v}), gradually removes symmetry elements – mirror planes (σ_d and σ_v), improper (S_4) and proper (C_3 and C_2) axes of rotation – in a very obvious manner. However, of more relevance to us in the present context is the incorporation of one or more labels that lower the molecular symmetry in a more subtle fashion such that we may monitor the reactivity, mechanism of formation, or molecular dynamics without unduly perturbing the character of the original system.

As an example, we note that differently labelled *ortho* or *meta* disubstituted benzenes have only a single molecular mirror plane (C_S symmetry) and are achiral; upon coordination of an organometallic moiety to one of the faces,[2] this mirror plane is broken, the molecular symmetry is reduced to C_1, and the molecule can exist in enantiomeric forms, as in Figure 1.2.

Likewise, when the individual ligands in octahedral metal complexes of the type $[M(NH_3)_6]^{x+}$ are linked in a pairwise fashion to form a system such as $[M(en)_3]^{x+}$, where *en* is ethylenediamine $H_2N(CH_2)_2NH_2$, this removes a number of symmetry elements — mirror planes (σ_h and σ_d), improper (S_6 and S_4) and proper (C_4, C_3 and C_2) axes of rotation, as well as the inversion centre (i); the symmetry is reduced to D_3 (only a single C_3 and three C_2's remain) and the molecule is chiral (Figure 1.3).

This was precisely the approach taken by Alfred Werner (Figure 1.4) when faced with the problem of elucidating the geometric structure of coordination complexes of the type ML_6, decades before the advent of X-ray crystallography as a standard technique for structural

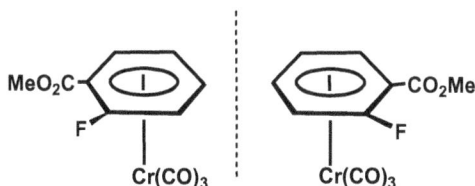

Figure 1.2. Enantiomers of (η^6-1-carbomethoxy-2-fluorobenzene)tricarbonylchromium(0).

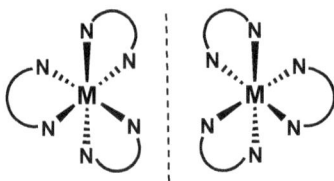

Figure 1.3. Enantiomeric tris(ethylenediamine)metal complexes, M(en)$_3$, of D_3 symmetry.

Figure 1.4. Alfred Werner (1866–1919) was a professor at the University of Zürich. He is considered the father of Coordination Chemistry (Nobel Prize 1913).

determination. Assuming that all six ligands were positioned equidistant from the central metal ion, he considered three possibilities: hexagonal planar, trigonal prismatic and octahedral. He first took the standard tactic of counting isomers: it is clear that complexes of the type MA$_5$X can have only a single isomer, whereas MA$_4$X$_2$ can potentially exist in three, three, and two forms for the hexagon, trigonal prism and octahedron, respectively. Likewise, the same result is predicted in the MA$_3$X$_3$ case (Figure 1.5).

Experimentally, however, the number of isomers found for MA$_5$X, MA$_4$X$_2$ and MA$_3$X$_3$ was one, two and two, respectively, suggesting the correct structure to be octahedral. Nevertheless, this was not totally definitive, since a third isomer might be unstable or difficult to isolate. Werner solved the problem brilliantly by preparing complexes of

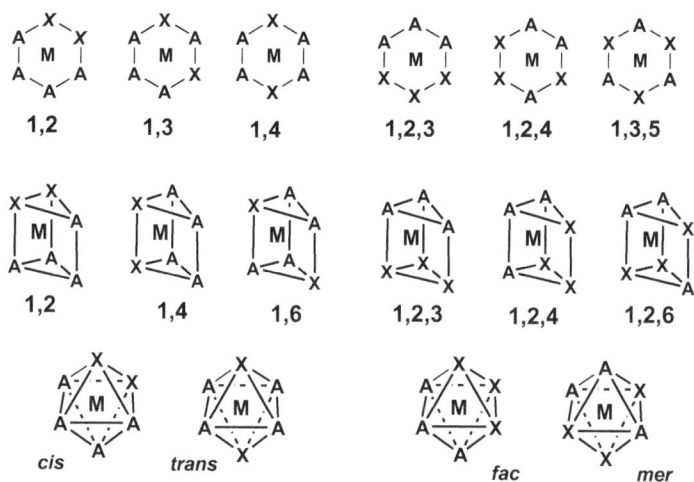

Figure 1.5. All possible MA$_4$X$_2$ and MA$_3$X$_3$ isomers derivable from six-coordinate hexagonal, trigonal prismatic and octahedral metal complexes.

Figure 1.6. All possible tris(ethylenediamine)metal complexes, M(*en*)$_3$, of hexagonal or trigonal prismatic geometry would be achiral.

the type M(A–A)$_3$, in which the bidentate ligands could only span adjacent sites, and he was able to separate them as enantiomers.[3] The corresponding structures for the hexagon and trigonal prism (Figure 1.6) all possess multiple mirror planes and are achiral.

The history and subsequent development of this work has been comprehensively reviewed.[4] We note in passing that the separation of diastereomeric [Co(*en*)$_3$]$^{3+}$ (+)-tartrate salts by fractional crystallisation is now routinely carried out in undergraduate laboratory classes worldwide. Subsequent replacement of the chiral tartrate ligands by chlorides furnishes the required enantiomers.

1.2. Reaction Mechanisms

The establishment of the mechanism of a chemical reaction can be a perilous endeavour, fraught with difficulties. It has frequently led to controversy and bitter disagreement, sometimes between Nobel Laureates. One must initially consider as many possibilities as can be envisaged, and then try to devise experiments that could distinguish between them. Ideally, such experiments should not merely support your favoured mechanism, but should also eliminate others. To quote a great physicist, Enrico Fermi, when discussing an experiment: *"There are two outcomes: if the result confirms the hypothesis, you have made a measurement. If the result is contrary to the hypothesis, you've made a discovery."*

The elucidation of reaction mechanisms has evolved greatly since the approaches developed by the early pioneers. Many now-classic experiments were carried out more than three-quarters of a century ago, but they demonstrated and verified techniques that have since been adopted, modified and improved by using currently available instrumentation. No doubt our own successors will add further enhancements. We shall remind ourselves about a number of these famous pioneering experiments, and then discuss some more recent advances, such as the detection and elucidation of otherwise hidden rearrangement processes, or reactions in which the detailed mechanism was obscure.

Many of those classic experiments focussed on the breaking of molecular symmetry, for example under thermolytic or photolytic conditions, or upon treatment with acids or bases. Numerous studies took advantage of the retention, inversion, or loss of chirality, the incorporation of isotopic labels such as carbon-14 or deuterium, as well as careful kinetic measurements. In many of the isotopic labelling experiments, it was necessary to bring about stepwise chemical degradation of the products, and then mentally reassemble the pieces to establish the structure of the rearranged molecule, and the location of the label. We discuss some of the more famous examples in Chapter 2.

In recent years, however, the remarkable advances in analytical instrumentation, such as Fourier transform infrared (FTIR) spectrometers with their immensely enhanced sensitivity, or the enormous increases in field/frequency in nuclear magnetic resonance (NMR) spectroscopy, have engendered new approaches. In the latter case, the revolutionary advances, moving from continuous wave (CW) NMR instruments, typically operating at 100 MHz, to superconducting Fourier-transform spectrometers, widely available at 600 MHz and above with their spectacular multi-pulse, multi-dimensional and multi-nuclear capabilities, have completely changed the experimental landscape. Typically, the use of ^{14}C isotopic labels with the requirement of radiochemical detection techniques has been largely superseded by ^{13}C NMR studies, whereby the progress of reactions can be conveniently followed continuously over a wide range of temperatures.

The ready availability and versatility of these new spectroscopic and spectrometric techniques[5-7] has prompted a multitude of mechanistic studies of a wide variety of novel reactions and molecular rearrangements in the organic, organometallic and inorganic chemistry domains. In particular, judicious symmetry breaking has facilitated the detection and elucidation of a large number of hidden rearrangement processes, many of which we describe in Chapter 5.

Other major advances have changed the way crystallography is now routinely utilised. The development of more powerful X-ray sources, such as rotating anodes, in conjunction with enhanced data collection systems (image plates, charge-coupled device area detectors, etc.) have, in very many cases, reduced data acquisition times from many days to just a few hours. Moreover, the newer direct methods programs,[8,9] combined with the continual improvements in computer technology, have reduced the time for structure solution such that X-ray crystallography can be used for the rapid identification of new materials, also furnishing the absolute configuration of chiral molecules. Nowadays, it is even possible to follow directly on the diffractometer the transformation of one single crystalline material into another.[10]

Perhaps the most valuable breakthrough in advancing our mechanistic understanding has been the development of the concept of the *Conservation of Orbital Symmetry*, primarily by Robert B. Woodward and Roald Hoffmann.[11] Among their seminal contributions, they not only rationalised many previously poorly understood molecular transformations, but also made numerous predictions, very many of which have since been validated experimentally. Frequently in such systems, astutely designed controlled symmetry breaking was a crucial component of their experimental verification, as we discuss in Chapter 3.

1.3. Synthetic Goals

Historically, the most challenging targets for the synthetic organic chemist have been natural products, generally those with biomedical activity, often with the aim of developing novel or improved routes to hitherto difficultly available molecules. One can list an enormous number of such cases, ranging in complexity from glucose by Emil Fischer,[12] to penicillin by Sheehan,[13] to vitamin B_{12} by Woodward[14] and Eschenmoser,[15] to taxol, brevetoxin B, and beyond by Nicolaou.[16] In those cases, the existence of the molecules in nature is self-evident, but their accessibility in viable quantities may be markedly problematic.

More recently, however, there has been a burgeoning number of research programmes focussed on the preparation of previously unknown molecules that exhibit unusual characteristics, such as very high symmetry, or the adoption of previously unforeseen bonding situations, such as quadruple or quintuple metal-metal bonds. These synthetic targets include molecules whose structures parallel those of the Platonic solids and of their derivatives, and those whose point group symmetry had not previously, or only rarely, been encountered. We shall discuss several of the synthetic brilliancies and mechanistic novelties arising from such systems.

1.4. New Reactions and Reagents

Although new reactions are continually being discovered, among the most important general synthetic procedures developed in recent years are alkene metathesis (Chauvin, Schrock, Grubbs),[17-19] and palladium-catalysed cross-coupling reactions (Heck, Negishi, Suzuki),[20-22] including those leading to the design and synthesis of molecular machines (Feringa, Sauvage, Stoddart),[23-25] all of which were deservedly recognised by the award of Nobel Prizes. Once again, a plethora of mechanistic proposals were advanced, and symmetry breaking played an important role in determining the now-accepted versions.

A number of exceedingly elegant contributions involving isotopic desymmetrisation have been made by synthetic chemists and have had major significance in the biochemical arena. The ready availability of such moieties as the chiral methyl group CHDT (Cornforth, Arigoni, Floss),[26-28] in which all three isotopes of hydrogen can be assembled in an enantiopure fashion, has revealed the stereochemical course of methyl migrations in the elucidation of biosynthetic pathways. Moreover, the isotopically chiral phosphate group, comprising $P[^{16}O^{17}O^{18}O]$ (Knowles, Lowe),[29,30] has been used to show how transphosphorylations can proceed in several stages, whereby successive inversions result in overall retention of chirality at phosphorus.

1.5. Additional Comments

We emphasise that our discussions of symmetry breaking are those cases in which the molecular symmetry has been deliberately lowered (but in such a way as to minimise any changes in molecular geometry or reactivity) thus permitting clarification of a reaction mechanism or rearrangement process. We are not focussing on those systems whereby the molecular symmetry is already intrinsically broken because of other factors. Typically, six-coordinate d^9-copper(II) complexes do not maintain their octahedral symmetry because the unequal population of the e_g orbitals brings about a Jahn-Teller distortion resulting in a D_{4h} structure with two long axial and four shorter equatorial distances.

Likewise, in 1,1-difluoroethylene the "apparently equivalent" fluorine nuclei give rise to an unexpectedly complex ^{19}F NMR spectrum because each fluorine couples differently to its neighbouring hydrogens, J_{H-F} (*trans*) $\neq J_{H-F}$ (*cis*); the fluorine nuclei are (as also are the hydrogens) magnetically non-equivalent and exhibit spin-spin coupling to each other. In contrast, in the corresponding allene $F_2C=C=CH_2$, the orthogonal relationship of the molecular termini

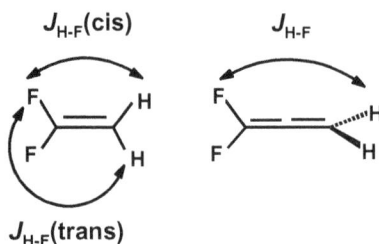

Figure 1.7. Coupling of fluorine to hydrogen is different in 1,1-difluoroethylene but the same in 1,1-difluoroallene.

places each fluorine equidistant from both hydrogens (Figure 1.7) resulting in a simple 1:2:1 triplet ^{19}F NMR spectrum.

Our objective here is to illustrate symmetry breaking and symmetry making in a broad range of reactions and molecular rearrangements in organic, inorganic and organometallic chemistry. Inevitably, these may to some extent be influenced by the author's own interests, but one can hope that they are balanced with appropriate examples selected from all across the Periodic Table.

References

1. R. Noyori and H. Takaya, BINAP – An efficient chiral element for asymmetric catalysis. *Acc. Chem. Res.* **1990**, *23*, 345–350.
2. L. Tchissambou, R. Dabard and G. Jaouen, Kinetic study on nucleophilic substitution and racemization in the benchrotrene series. *C.R. Acad. Sci. Ser. C.* **1972**, *274*, 806.
3. A. Werner, Concerning the asymmetric cobalt atom. *Chem. Ber.* **1912**, *45*, 121–130.
4. A. Ehnbom, S.K. Ghosh, K.G. Lewis and J.A. Gladysz, Octahedral Werner complexes with substituted ethylenediamine ligands: a stereochemical primer for a historic series of compounds now emerging as a modern family of catalysts. *Chem. Soc. Rev.* **2016**, *45*, 6799–6811.
5. R.R. Ernst, Nuclear magnetic resonance Fourier-transform spectroscopy (Nobel Lecture). *Angew. Chem. Int. Ed.* **1992**, *31*, 805–823.
6. J.B. Fenn, Electrospray wings for molecular elephants (Nobel Lecture). *Angew. Chem. Int. Ed.* **2003**, *42*, 3871–3894.
7. K. Tanaka, The origin of macromolecule ionization by laser irradiation (Nobel Lecture). *Angew. Chem. Int. Ed.* **2003**, *42*, 3860–3870.

8. H. Hauptman, Direct methods and anomalous dispersion (Nobel Lecture). *Angew. Chem. Int. Ed.* **1986**, *25*, 603–613.

9. J. Karle, Recovering phase information from intensity data (Nobel Lecture). *Angew. Chem. Int. Ed.* **1986**, *25*, 614–629.

10. B.B. Rath and J.J. Vittal, Single-crystal-to-single-crystal [2+2] photocycloaddition reaction in a photosalient one-dimensional coordination polymer of Pb(II). *J. Am. Chem. Soc.* **2020**, *142*, 20117–20123, and references therein.

11. R.B. Woodward and R. Hoffmann, The conservation of orbital symmetry. *Angew. Chem. Int. Ed.* **1969**, *8*, 781–853.

12. E. Fischer, Synthesis of glucose. *Ber. Dtsch. Chem. Ges.* **1890**, *23*, 799–805.

13. J.C. Sheehan and K.R. Henery-Logan, A general synthesis of the penicillins. *J. Am. Chem. Soc.* **1959**, *81*, 5838–5839.

14. R.B. Woodward, Total synthesis of Vitamin B_{12}. *Pure Appl. Chem.* **1971**, *25*, 283–306.

15. A. Eschenmoser, Organic natural product syntheses today: Vitamin B_{12} as an example. *Naturwiss.* **1976**, *61*, 513–525.

16. K.C. Nicolaou, Organic synthesis: the art and science of replicating the molecules of living nature and creating others like them in the laboratory. *Proc. Roy. Soc. A* **2014**, *470*, 20130690.

17. Y. Chauvin, Olefin metathesis: the early days (Nobel Lecture). *Angew. Chem. Int. Ed.* **2006**, *45*, 3741–3747.

18. R.R. Schrock, Multiple metal-carbon bonds for catalysis reactions (Nobel Lecture). *Angew. Chem. Int. Ed.* **2006**, *45*, 3748–3759.

19. R.H. Grubbs, Olefin metathesis catalysts for the preparation of molecules and materials (Nobel Lecture). *Angew. Chem. Int. Ed.* **2006**, *45*, 3760–3765.

20. R.F. Heck, Aromatic haloethylation with palladium and copper halides. *J. Am. Chem. Soc.* **1968**, *90*, 5535–5538.

21. E-i. Negishi, Magical power of transition metals: past, present and future (Nobel Lecture). *Angew. Chem. Int. Ed.* **2011**, *50*, 6738–6764.

22. A. Suzuki, Cross-coupling reactions of organoboranes: an easy way to construct C-C bonds (Nobel Lecture). *Angew. Chem. Int. Ed.* **2011**, *50*, 6722–6737.

23. B.L. Feringa, The art of building small: from molecular switches to motors (Nobel Lecture). *Angew. Chem. Int. Ed.* **2017**, *56*, 11060–11078.

24. J-P. Sauvage, From chemical topology to molecular machines (Nobel Lecture). *Angew. Chem. Int. Ed.* **2017**, *56*, 11080–11093.

25. J.F. Stoddart, Mechanically interlocked molecules (MIMs) – molecular shuttles, switches and machines (Nobel Lecture). *Angew. Chem. Int. Ed.* **2006**, *56*, 11094–11125.

26. J.W. Cornforth, J.W. Redmond, H. Eggerer, W. Buckel and C. Gutschow, Asymmetric methyl groups. *Nature* **1969**, *221*, 1212–1213.

27. J. Lüthy, J. Rétey and D. Arigoni, Preparation and detection of chiral methyl groups. *Nature* **1969**, *221*, 1213–1215.

28. H.G. Floss and S. Lee, Chiral methyl groups: small is beautiful. *Acc. Chem. Res.* **1993**, *26*, 116–122.

29. S.J. Abbott, S.R. Jones, S.A. Weinman and J.R. Knowles, Chiral [^{16}O,^{17}O,^{18}O]-phosphate monoesters. 1. Asymmetric synthesis and stereochemical analysis of [1(R)-^{16}O,^{17}O,^{18}O]phospho-(S)-propane-1,2-diol. *J. Am. Chem. Soc.* **1978**, *100*, 2558–2560.

30. G. Lowe, Chiral [^{16}O,^{17}O,^{18}O]phosphate esters. *Acc. Chem. Res.* **1983**, *16*, 244–251.

Chapter 2

Symmetry Breaking in Classic Mechanistic Investigations

"I was taught that the way of progress was neither swift nor easy."
— *Marie Curie*

2.1. Benzyne

A beautiful example of the power of symmetry breaking to elucidate a reaction mechanism was the demonstration of the existence of benzyne as a reaction intermediate. Treatment of chlorobenzene with potassium amide in liquid ammonia to form aniline could be envisaged as a direct nucleophilic displacement of chloride, or by elimination of HCl to form an intermediate dehydrobenzene, C_6H_4, containing a triple bond. Since nucleophilic attack on an arene ring normally requires the presence of strongly electron-withdrawing substituents such as nitro or cyano, J.D. Roberts, then at MIT and later at Caltech, chose to seek evidence for the elimination-addition sequence by preparing chlorobenzene in which the *ipso* carbon was labelled with carbon-14. When the reaction was carried out, the resulting aniline was found to bear approximately 50% of the radiolabel at the *ipso* carbon, with the remainder at the positions *ortho* to it, as shown in Scheme 2.1.[1]

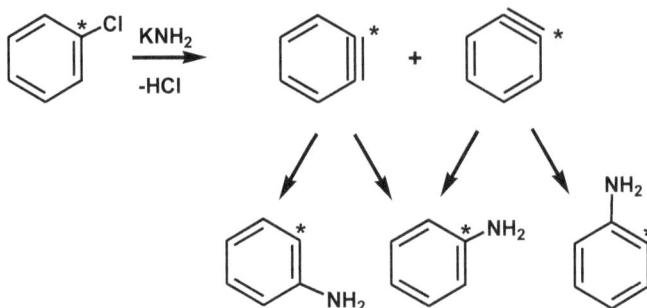

Scheme 2.1. Dehydrochlorination of chlorobenzene to form aniline via benzyne.

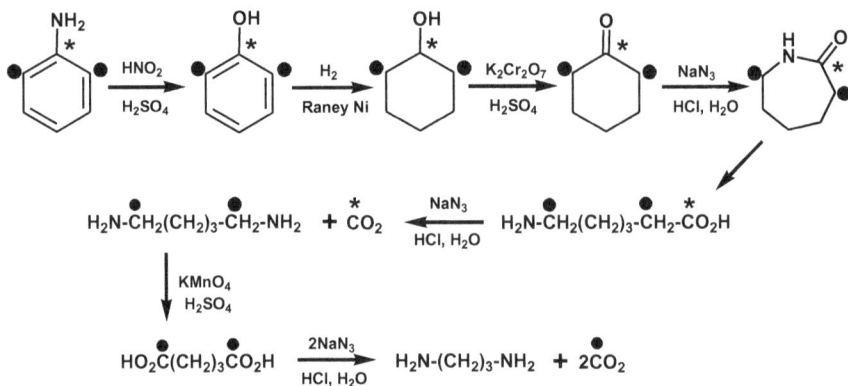

Scheme 2.2. Chemical degradation of ^{14}C-labelled aniline to release CO_2.

If the ^{14}C label had been found only in the carbon attached to nitrogen, this would have indicated a direct substitution process, whereas the observed result clearly indicated the formation of an intermediate benzyne in which the triple-bonded carbons had become equivalent thus allowing nucleophilic attack at either position.

While this may appear to be a simple elegant procedure, experimentally it was by no means a trivial endeavour. First of all, only a small percentage of the molecules were radioactively labelled, and secondly, it was necessary to locate their positions and relative abundances in the final product. As shown in Scheme 2.2, in the present case, the ^{14}C-labelled aniline was treated initially with nitrous acid to form the diazonium salt that, upon hydrolysis, gave phenol. Catalytic hydrogenation to cyclohexanol, oxidation to cyclohexanone, and reaction with hydrazoic acid brought about a Schmidt rearrangement to caprolactam that was hydrolysed to form 6-aminohexanoic acid. Further treatment with HN_3 to produce a carbamic acid yielded 1,5-diaminopentane, and liberated the first unit of CO_2. Subsequent oxidation to give glutaric acid, and yet another reaction with HN_3 furnished 1,3-diaminopropane and the final portion of CO_2. In each case, the evolved carbon dioxide was trapped initially in NaOH, and its ^{14}C content finally recorded as barium carbonate.

If the original ammonolysis had proceeded by direct replacement of chloride, all the ^{14}C label would have been in the keto group of the cyclohexanone, and would have been found in the CO_2 initially released; in fact, only half was found there, and the remainder was obtained from the glutaric acid upon further reaction with hydrazoic acid. While one can only admire the skill and dedication of these experimenters, we note that, with today's instrumentation, such an experiment would be relatively straightforward by labelling with carbon-13 and assigning the markedly enhanced resonances in the ^{13}C NMR spectrum.

Since that time, a multitude of arynes, including heterocycles such as pyridyne from pyridine, have been prepared and their reactivity extensively investigated.[2] Of particular importance have been the reactions of benzyne, now frequently generated by elimination of CO_2 and N_2 from benzenediazonium-2-carboxylate, and its derivatives in Diels-Alder cycloadditions thus giving rise, for example, to barrelenes and triptycenes.

2.2. The Favorskii Rearrangement

In the Favorskii rearrangement (Figure 2.1), the treatment of α-haloketones with alkoxides yields esters; in the particular case of cyclic systems, the resulting esters suffer ring contraction. Since it was well known that certain isomeric substrates, such as 1 and 2 exemplified in Scheme 2.3, yielded the same rearrangement products, it was suspected that the reaction involved a symmetrical intermediate.[3] The proposed mechanism required initial deprotonation at a carbon adjacent to the carbonyl group, intramolecular carbanionic displacement

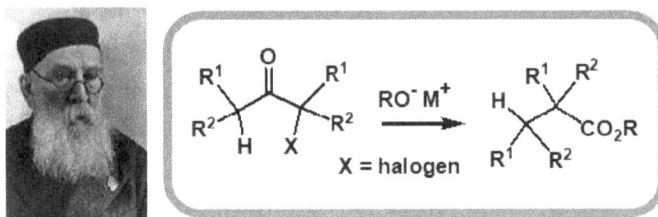

Figure 2.1. Alexei Favorskii (1860–1945), a professor at St. Petersburg University in Russia, first reported this rearrangement in *J. Russ. Phys. Chem. Soc.* **1894**, *26*, 530.

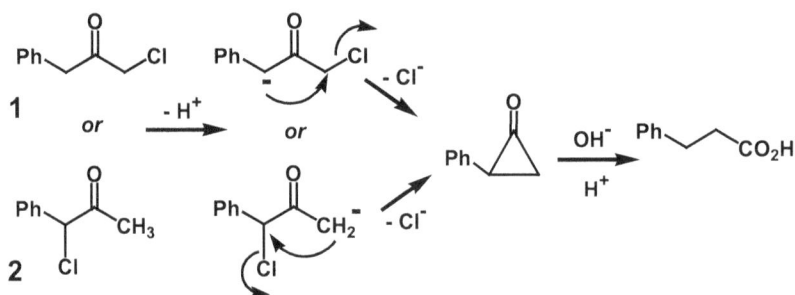

Scheme 2.3. Favorskii rearrangement proceeds via ring opening of a cyclopropanone.

of the halide to form a cyclopropanone, followed by opening of the strained ring by alkoxide, or hydroxide, to furnish the observed ester, or carboxylic acid.

This was unequivocally and elegantly verified by Robert Loftfield at Harvard who treated doubly ^{14}C-labelled 2-chlorocyclohexanone, **3**, with sodium isoamyloxide and located the labels in the isoamyl cyclopentane carboxylate, as illustrated in Scheme 2.4. It was found that the level of radioactivity was maintained on the carbonyl carbon and the remainder was equally split between C(1) and C(2) of the cyclopentane ring. This result clearly verified the intermediacy of the cyclopropanone, **4**, whereby the ring opening could occur with equal probability in two ways, thus splitting the distribution of the radiolabel.[4]

The Favorskii rearrangement has been widely used in synthesis, notably in the first reported preparation of cubane, by Eaton and

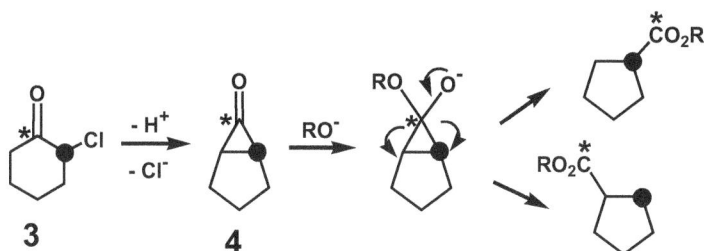

Scheme 2.4. Loftfield's carbon-14 double-labelling experiment to verify the mechanism.

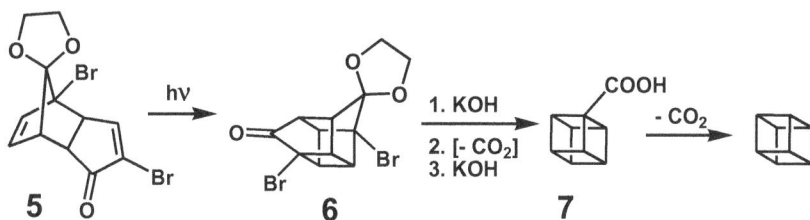

Scheme 2.5. The first synthesis of cubane by Eaton and Cole.

Cole, in which it is used twice.[5] As shown in Scheme 2.5, photolysis of **5**, the monoketal of the Diels-Alder dimer of 2-bromo-cyclopentadienone, yields **6**, which is exquisitely poised to undergo successive ring contractions to yield eventually the carboxylic acid, **7**, and ultimately cubane, the desired Platonic solid.

2.3. On the Mechanism of the Walden Inversion

It is now accepted dogma that S_N2 reactions proceed via Walden (Figure 2.2) inversion at the carbon undergoing nucleophilic attack. In a now celebrated experiment, E.D. Hughes from University College London invoked not just one, but two, changes in symmetry. When an enantiopure alkyl iodide, such as $2R$-iodooctane, was added to a solution containing free iodide ions, the alkyl halide underwent racemisation at a rate conveniently followed by polarimetry. More ingeniously however, as depicted in Scheme 2.6, when the incoming

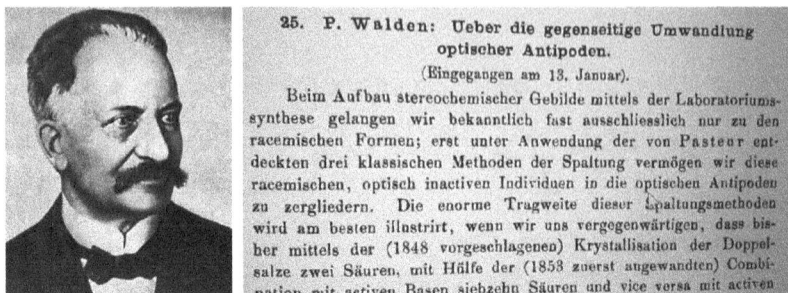

Figure 2.2. Paul Walden (1863–1957) served as Head of the Chemistry Department at the University of Rostock from 1919 to 1934, and published a famous paper "Concerning the mutual conversion of optical antipodes" in *Ber. Dtsch. Chem. Ges.* **1896**, *29*, 133–138.

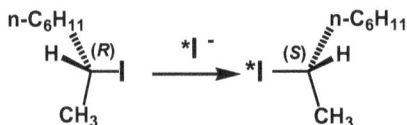

Scheme 2.6. Walden inversion of 2-iodooctane using isotopically labelled iodide.

iodide was labelled as ^{128}I, the rate of racemisation was found to be twice the rate of incorporation of the radioactive iodine.[6] This must be so because for a given group of alkyl iodide molecules with configuration *R*, racemisation will be complete when just half of them have been inverted, but only 50% of them will be radiolabelled.

If the reaction is taken too far, there are minor complications attributable to the reversibility of the reaction, and also from slow decay of the radioisotope. Nevertheless, the case for complete inversion in all S_N2 processes was unequivocally demonstrated. Moreover, this work was subsequently extended to studies using radiolabelled bromide in reactions with enantiopure α-phenethyl bromide or α-bromopropionic acid, and identical results were obtained.[7,8]

While the phenomenon of inversion in S_N2 displacements has long been established, the verification of a linear alignment of incoming nucleophile and leaving group is of much more recent vintage.[9] Treatment of methyl α-tosyl-*o*-toluenesulfonate, **8**, with base gave the

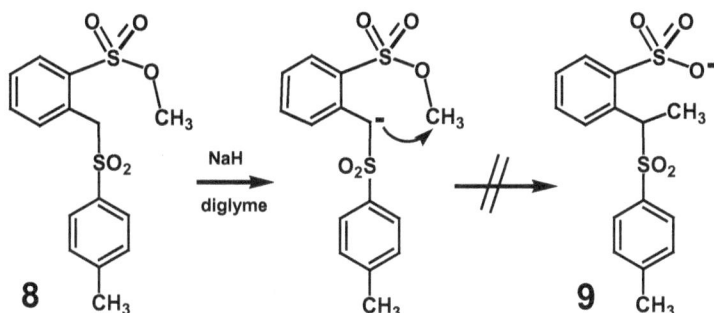

Scheme 2.7. The hypothetical, but unsuccessful, intramolecular S_N2 transfer of a methyl group from oxygen to carbon.

o-(1-tosylethyl)benzenesulfonate ion, **9**, and one might naively have envisaged an intramolecular S_N2 reaction, thus bringing about a methyl transfer from oxygen to carbon, as in Scheme 2.7. However, this is not what occurs, and in a beautifully designed crossover experiment from Eschenmoser's laboratory at ETH Zürich, it was established that the reaction actually proceeds in an intermolecular fashion and provides crucial additional insight into the S_N2 mechanism.

Having synthesised the starting material in which both methyl groups were fully deuterated, an equimolar mixture of d_0-**8** and d_6-**8** was deprotonated, and the products analysed via mass spectrometry. The striking result of this experiment was that the products were found to be d_0-**9**, d_3-**9** and d_6-**9** in an approximate 1:2:1 ratio, as depicted in Scheme 2.8.[9] An exclusively intramolecular process would yield only d_0-**9** and d_6-**9** products, but the observed ratio clearly indicates that the intermediate carbanion always attacks a neighbouring molecule because, in the intramolecular case, it cannot become aligned in the required linear transition state. This result contrasts markedly with those cases in which there are no geometric constraints, and intramolecular S_N2 reactions can proceed freely.

We note in passing that one might have predicted complete racemisation of a chiral substrate in S_N1 reactions, since the intermediate trigonal planar $[CRR'R'']^+$ species can be attacked from either side, but this is generally not the case because of complications such as tight ion pair formation, or solvent cage effects.[10]

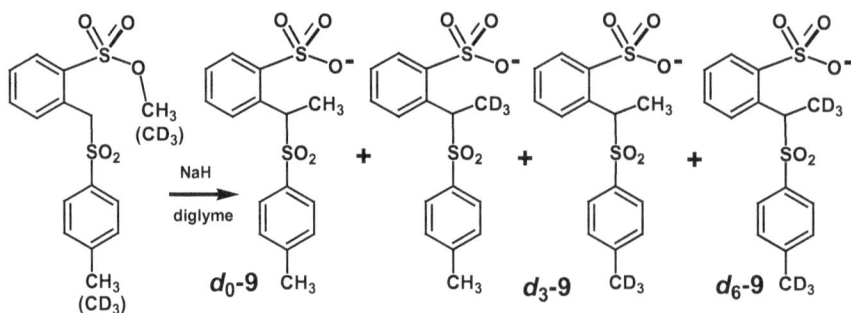

Scheme 2.8. An intermolecular S_N2 process leads to scrambling of the trideuteromethyl groups.

2.4. Migration onto an Electron-deficient Carbon Centre

2.4.1. *The Pinacol-pinacolone Rearrangement*

The rearrangement of pinacols to pinacolones was initially discovered by Fittig (Figure 2.3) in 1859,[11] but was only correctly interpreted by Butlerov (Figure 2.4) in 1870,[12] who also used this route to prepare trimethylacetic acid.

The pinacol-pinacolone rearrangement has been used to study migratory aptitude,[13] a widely invoked parameter to account for the selectivity and ease of molecular rearrangements. Starting from a symmetric precursor, one can see which of several substituents can most readily migrate. Typically, in cationic species, it is commonly assumed that a phenyl migration is favoured over that of its methyl counterpart because of charge delocalisation in the phenonium ion transition state. However, such generalisations must be treated with caution. For example, upon protonation of the pinacol $Ph_2C(OH)–CMe_2OH$ it is a methyl that preferentially migrates (Scheme 2.9), not because the alkyl group has a better migratory aptitude than the aryl substituent, but rather that the more favoured carbocation, **10**, is doubly benzylic.[14]

To make a fair comparison, such as between phenyl and ferrocenyl, one must eliminate these extraneous factors, for example, by placing the competing substituents on either side of a molecular

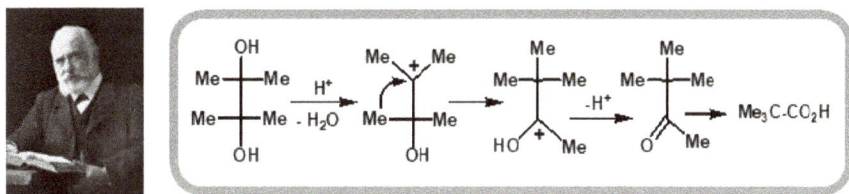

Figure 2.3. Wilhelm Rudolf Fittig (1835–1910) worked at the Universities of Göttingen and Tübingen; he prepared 1,2-diols (pinacols) from ketones and investigated their behaviour.

Alexander Butlerov (1828-1886) of the University of St. Petersburg correctly recognized the products of the pinacol-pinacolone rearrangement, and their significance with respect to the preparation of trimethylacetic acid. His life was celebrated by the issue of a Russian postage stamp. Favorskii was one of his doctoral students.

Figure 2.4. Alexander Butlerov.

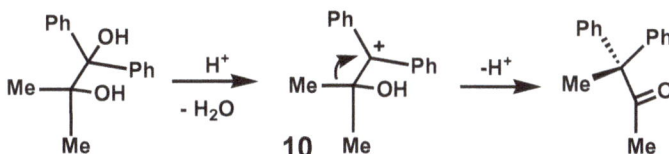

Scheme 2.9. Pinacol rearrangement showing methyl rather than phenyl migration.

mirror plane. This was first discussed by Weliky and Gould in 1956 who prepared 1,2-diferrocenyl-1,2-diphenyl-ethanediol, **11**, by treatment of benzoylferrocene with cobalt dichloride in the presence of a Grignard reagent.[15] They noted that this pinacol *"rearranged with remarkable ease (merely upon passing gaseous HCl over its solution in benzene) to the pinacolone* **12**". Clearly, ferrocenyl was favoured over phenyl migration, as in Scheme 2.10. Later work by several groups[16-18] revealed, not surprisingly, that ferrocenyl migration was also favoured over an alkyl shift (R = Me, Et, ^{n}Pr, ^{n}Bu). Subsequently, however, it

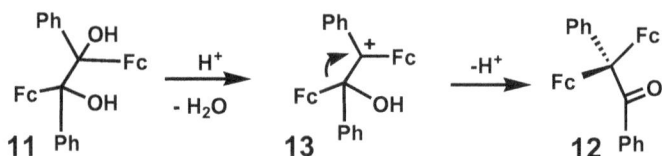

Scheme 2.10. Pinacol rearrangement showing favoured migration of ferrocenyl over phenyl.

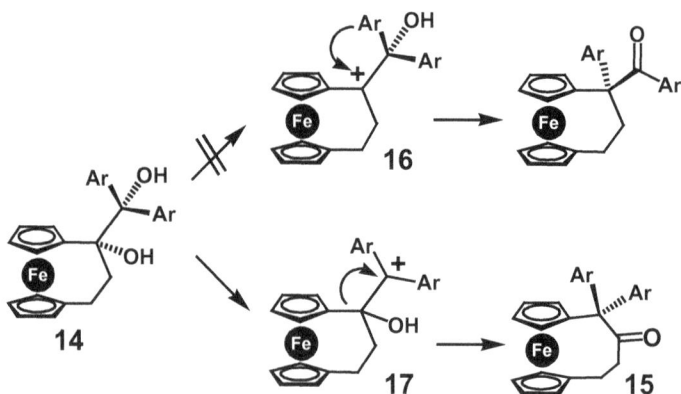

Scheme 2.11. Carbocation formation at a doubly benzylic site, rather than α to *ansa*-ferrocenyl.

was found that phenyl migration in cation **13** is indeed observable, but that the ferrocenyl shift was preferred by a factor of ~8.[19]

In a more recent example involving the ferrocenyl moiety, protonation of the pinacol, **14**, arising from McMurry coupling of [3]ferrocenophanone and benzophenone, furnished only the pinacolone, **15**, arising from ferrocenyl rather than phenyl migration (Scheme 2.11).[20] In this system, a number of factors come into play: one would generally expect carbocation formation at the position α to the ferrocenyl group, as in **16**, rather than the benzylic cation, **17**, since the sandwich moiety is known to stabilise such species via delocalisation of positive charge.[21] However, in this relatively rigid, strained system the potential ferrocenyl-stabilised cation, **16**, cannot gain the advantage of leaning towards the metal atom so as to optimise orbital overlap between the vacant $2p$ orbital on carbon with a filled $3d$ orbital on iron. Moreover, in the observed product there is relief of ring strain whereby the bridge linking the cyclopentadienyl rings has now expanded to form the [4]ferrocenophanone, **15**.

2.4.2. *The Wolff Rearrangement*

In the Wolff rearrangement (Figure 2.5) α-diazoketones are converted, either thermally or photochemically, into ketenes that then react with nucleophiles, such as water or alcohols, to form carboxylic acids or esters. The reaction has been the subject of much debate as to whether migration occurs in a concerted fashion, or in a stepwise manner via a carbene that could adopt either a singlet or triplet spin state. The Wolff rearrangement and its remarkable versatility as a synthetic procedure has been very comprehensively reviewed by Kirmse.[22]

In terms of the mechanism, of particular interest was the potential intermediacy of an oxirene whose existence would facilitate carbene-to-carbene rearrangements. Once again, isotope labelling experiments were invoked, as depicted in Scheme 2.12. Photolysis of 2-diazobutan-3-one, **18**, forms dimethylketene that, upon hydrolysis with water, yields 2-methylpropionic acid. However, when labelled at the carbonyl position (initially with ^{14}C, later with ^{13}C), this reaction yielded ketenes in which the label was scrambled over the carbonyl and α-carbons, as seen in the carboxylic acid product;[23] of course, without a label these molecules are indistinguishable. The process can be rationalised in terms of an intermediate oxirene, **19**, through which carbene-carbene interchange can occur.

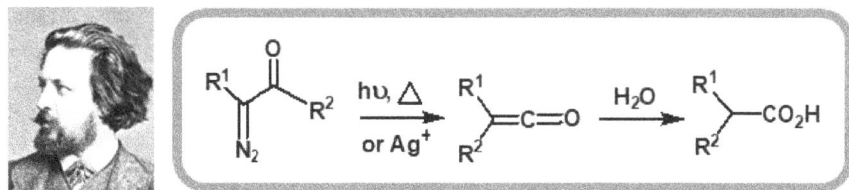

Figure 2.5.   Ludwig Wolff (1857–1919) was Professor of Analytical Chemistry at the University of Jena. He earned his doctorate in Fittig's laboratory.

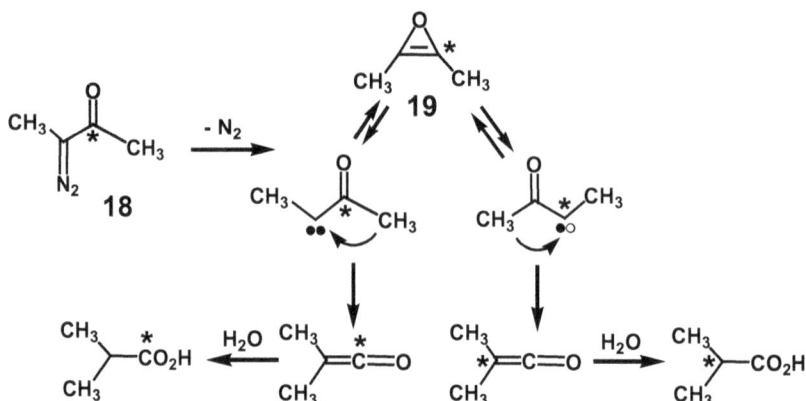

Scheme 2.12. A carbon-13 labelling experiment to demonstrate the intermediacy of an oxirene intermediate in a Wolff rearrangement.

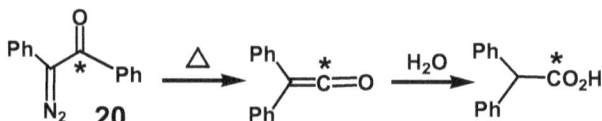

Scheme 2.13. Thermolysis of diazobenzil does not exhibit ^{13}C scrambling.

It is noteworthy that thermally induced Wolff rearrangement of the α-diazo-benzil, **20**, furnishes diphenylacetic acid in which the ^{13}C label remains on the carbonyl carbon indicating that the process occurs in a concerted manner (Scheme 2.13).[24]

2.5. Neighbouring Group Participation

The phenomenon of cationic charge alleviation by a neighbouring polarisable heteroatom has long been recognised. The very facile hydrolysis of CH_3CH_2–S–CH_2CH_2Cl, involving the intermediacy of a bridged sulfonium species (Scheme 2.14), is paralleled by the use of $(ClCH_2CH_2)_2S$ as a component of "Mustard Gas" that was used in wartime to release hydrogen chloride, to the enormous detriment of its recipients.

Building on this, it was recognised that the course of many reactions, especially rearrangements, are controlled by the maintenance,

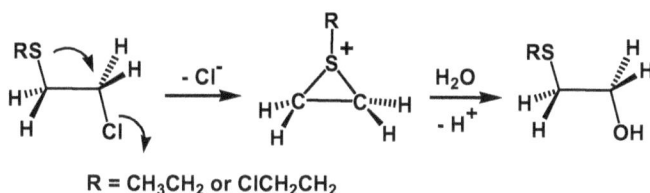

R = CH$_3$CH$_2$ or ClCH$_2$CH$_2$

Scheme 2.14. Hydrolysis via a bridged sulfonium species.

or loss, of their intrinsic molecular symmetry. In a series of important pioneering experiments, Saul Winstein then at Caltech studied the influence of neighbouring group participation to probe its stereochemical consequences by taking advantage of the subtle differences in behaviour between *erythro* and *threo* diastereomers.

> The descriptors *erythro* and *threo* are used to identify diastereomers containing two adjacent chiral centres bearing two pairs of common substituents, but the third is different. They are derived from the configurations of the sugars erythrose and threose. In *erythro* (*meso-like*) molecules, visualising them in an eclipsed conformation would align two pairs of identical substituents, whereas in the corresponding *threo* isomer they would lie on opposite sides.

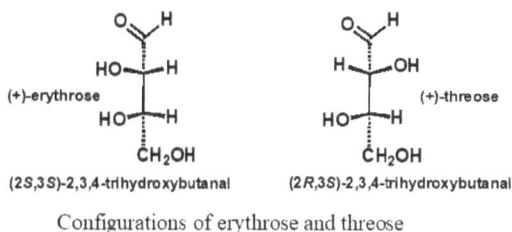

(+)-erythrose

(2S,3S)-2,3,4-trihydroxybutanal

(+)-threose

(2R,3S)-2,3,4-trihydroxybutanal

Configurations of erythrose and threose

As shown in Scheme 2.15, base-mediated elimination of HBr from racemic *threo* and *erythro* 3-bromobutan-2-ols, **21** and **22**, respectively, yields isomeric epoxides. In the former case, the antiperiplanar alignment of the incoming oxygen nucleophile and the bromide leaving group leads to the *cis* epoxide, **23**, a *meso* structure with C_S (mirror) symmetry and is thus achiral. In contrast, the *erythro* precursor yields the *trans* epoxide, **24**, of C_2 (chiral) symmetry, thereby generating a pair of enantiomers.[25]

However, in these cases, the positioning of the hydroxyl and bromine substituents on adjacent carbons, together with their known

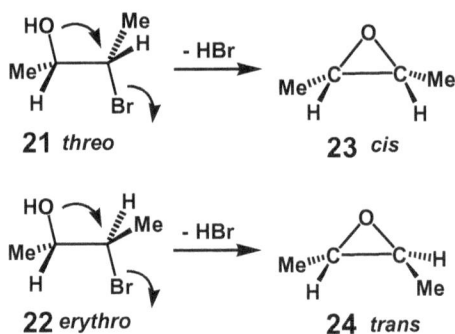

Scheme 2.15. *Threo* and *erythro* bromohydrins yield isomeric epoxides.

antiperiplanar alignment in the elimination reaction, made these results predictable. In contrast, the reaction of HBr with the optically pure *threo* and *erythro* diastereomers of 3-bromobutan-2-ols had more significant consequences – the unequivocal demonstration of neighbouring group participation (anchimeric assistance). In short, in both cases substitution of the hydroxyl group by bromine occurred with *retention* of the original relative stereochemistry, thereby indicating the intermediacy of a bromonium ion.[26]

Considering first the protonation of (2R,3R)-*threo*-3-bromobutan-2-ol, **25**, the positive charge developing at C(2) is partially alleviated by the neighbouring bromine that adopts a bridging position so as to generate a C_S (mirror-symmetric) intermediate, **26**, depicted in Scheme 2.16. Subsequent attack by bromide is equally probable at C(2) and C(3) and leads to a pair of enantiomeric dibromides.

In the 2S,3R-*erythro* isomer, **27**, the intermediate is now the C_2-symmetric bromonium species, **28**, that can suffer bromide addition at either carbon, giving in both cases the *meso* dibromide. Attack at the original C(2) hydroxyl-bearing carbon pushes the bridging bromine back to C(3), whereas addition at C(3) forces the original bromine to migrate; however, these are indistinguishable and lead to the same result. As illustrated in Schemes 2.16 and 2.17, overall retention is the result of a double inversion process, the first of which is intramolecular leading to bromonium ion formation.

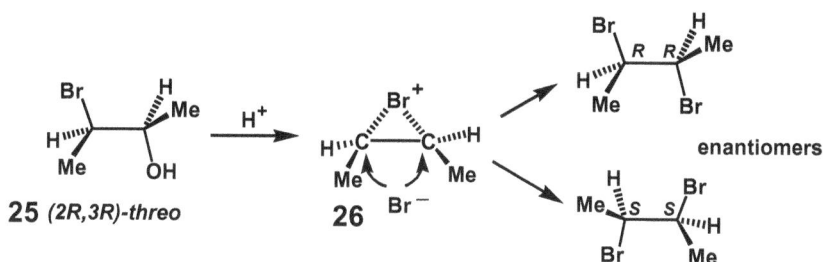

Scheme 2.16. The *threo* bromohydrin, **25**, yields a pair of enantiomeric *threo* dibromides.

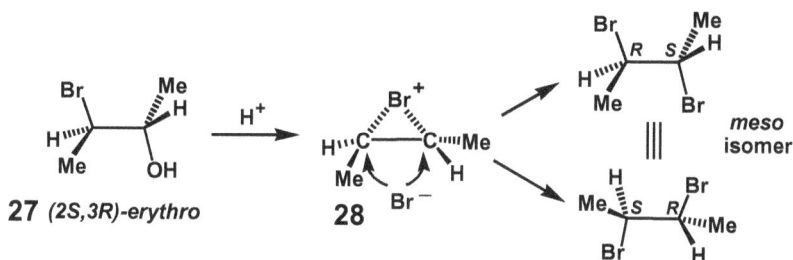

Scheme 2.17. The *erythro* bromohydrin, **27**, yields only the *meso* dibromide.

In the *threo* case, the mirror-symmetric bromonium ion, **26**, has carbons with R and S configurations. Approach by the nucleophile brings about inversion at the site of attack but leads to retention at the other carbon; hence, both now have the same configuration, (R,R) or (S,S), resulting in a pair of enantiomers. On the other hand, in the C_2-symmetric bromonium ion, both carbons have R configurations and nucleophilic attack inverts *only one of them* resulting in the (R,S) or (S,R) *meso* product.

As shown in Scheme 2.18, this approach was subsequently extended by Donald Cram (Nobel Prize in 1987 for work on host-guest chemistry) at UCLA to demonstrate the existence of phenonium ion intermediates whereby a phenyl ring bridges two carbons and is oriented orthogonal to the bond between them. In this work, optically pure *threo* and *erythro* diastereomers of the tosylate esters of

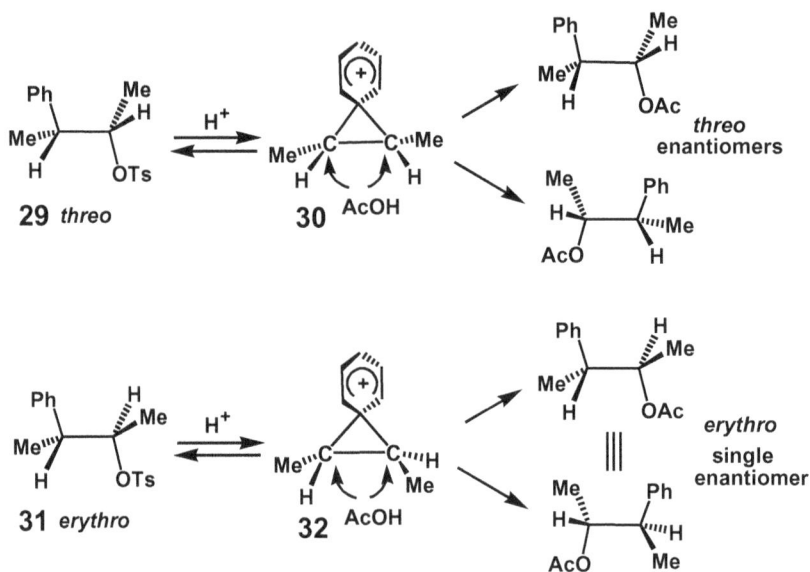

Scheme 2.18. Solvolysis of *threo* and *erythro* tosylates yields isomeric acetates.

3-phenylbutan-2-ol were solvolysed in acetic acid. The *threo* isomer, **29**, yielded the C_s-symmetric phenonium ion, **30**, whose *meso* character led to an enantiomeric pair of *threo* acetates. In contrast, the *erythro* isomer, **31**, yielded the C_2-symmetric phenonium ion, **32**, that furnished a single enantiomer of *erythro* acetate. We can clearly see how the existence of the mirror plane in phenonium ion **30** inevitably led to racemisation, whereas the C_2 symmetry of **32** preserved its chiral character.[27]

2.6. Migration onto an Electron-deficient Nitrogen Centre

2.6.1. *The Beckmann Rearrangement*

When oximes are treated with reagents (e.g., H_2SO_4, PCl_5, or $SOCl_2$) that convert the hydroxyl into a leaving group, it brings about migration of the *anti* substituent to form a nitrilium ion; subsequent reaction with water yields a secondary amide.[28] This specific geometric requirement contrasts with the situation in the pinacol rearrangement

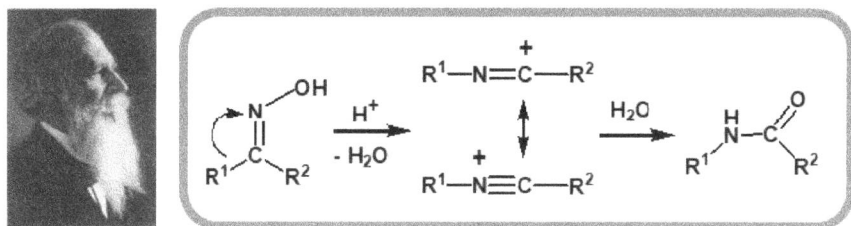

Figure 2.6. Ernst Otto Beckmann (1853–1923) spent much of his career at the University of Leipzig, working mostly with Kolbe and Ostwald.

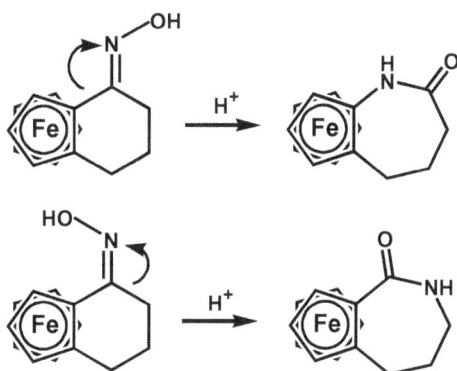

Scheme 2.19. Beckmann migration of *anti* and *syn* oximes of ferrocenylcyclohexanone.

whereby the identity of the migrating group is controlled primarily by its migratory aptitude rather than its geometric alignment. The Beckmann rearrangement (Figure 2.6) proceeds with retention, as was shown when the optically active $(PhCH_2)(CH_3)CH$ moiety was installed as the migrating group.[29]

A particularly fine example illustrating the geometric requirement for migration of the *anti* substituent is provided by the two oximes shown in Scheme 2.19 that rearrange to form isomeric ferrocenocaprolactams.[30]

2.6.2 *The Curtius Rearrangement*

Thermolysis of acyl azides proceeds with loss of nitrogen to form isocyanates. Although they can be isolated, the reaction is often carried out in the presence of a nucleophilic solvent such as an alcohol

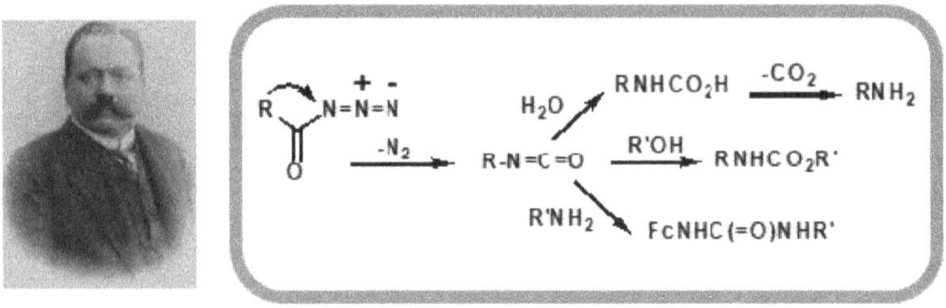

Figure 2.7. Theodor Curtius (1857–1928) worked initially with Bunsen, later with Kolbe at Leipzig, and eventually became Professor of Chemistry in Heidelberg.

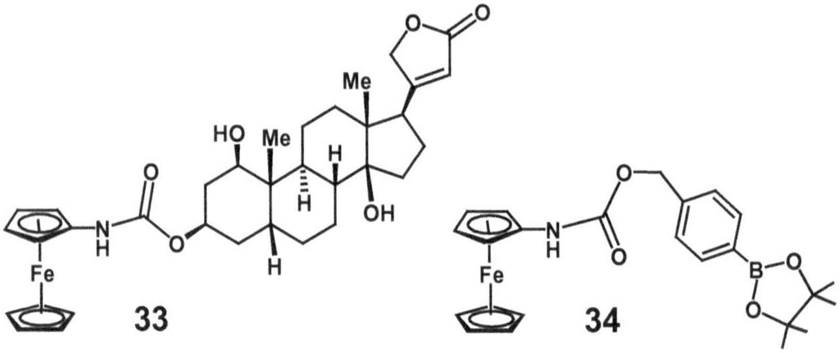

Figure 2.8. Steroidal and boronic acid ester carbamates formed by reaction of alcohols with ferrocenyl isocyanate.

or amine to form esters or ureas, or with water to form carbamic acids that lose carbon dioxide to form amines.[31] In contrast to the Wolff rearrangement of α-diazoketones that can proceed via a carbene intermediate, the Curtius rearrangement (Figure 2.7) is almost certainly a concerted process that does not involve a discrete nitrene. As with the Beckmann, the rearrangement proceeds with retention of the migrating group.[29,32]

The rearrangement of an acyl azide to form the corresponding amine was used by R.B. Woodward in the first synthesis of triquinacene (see Chapter 7),[33] and numerous practical applications of this procedure have since been reported. For example, Curtius rearrangement of ferrocenoyl azide in the presence of a number of hydroxysteroids to form carbamates, such as **33** (Figure 2.8), has been used in high performance liquid chromatography using electrochemical detection to characterise the products of the bioconversion of digoxigenin.[34] Moreover, it has been shown that molecules in which the ferrocenyl carbamate unit is linked to a boronic acid ester, as in **34**, provide a very sensitive and rapid electrochemical detection technique for hydrogen peroxide and glucose; this can be used to follow enzyme activity and cell signalling pathways.[35]

2.7. The Benzidine Rearrangement

Protonation of hydrazobenzene brings about cleavage of the HN–NH linkage and coupling of the phenyl substituents, normally in the *para* positions to form 4,4′-diaminobiphenyl (benzidine). For many years the mechanism attracted considerable controversy,[36] but much has since been clarified. In particular, the failure to observe mixed products in crossover experiments using differently substituted aryl groups definitively ruled out a dissociative mechanism. Crucial information from kinetic isotope studies in which both nitrogens of starting material were ^{15}N-labelled, or with a ^{14}C incorporated at a *para* position, indicated that formation of the new C–C linkage and cleavage of the N–N bond both take place in the rate-determining step.[37] The mechanism is therefore a concerted process, and can be described as a [5,5]-sigmatropic rearrangement, as depicted in Scheme 2.20.

2.8. Closing Remarks

In these historically important studies, the experimental methods used were of the now-classic type that required notable laboratory skills, and frequently involved further chemical manipulation of the products to secure their unambiguous identification. More recently, researchers have been able to benefit from the advent of a wide range of spectroscopic, spectrometric and crystallographic techniques by which the detailed structures of molecules could be determined directly on smaller quantities, and without recourse to further experimental modification. We shall describe and exemplify these approaches in the coming few chapters.

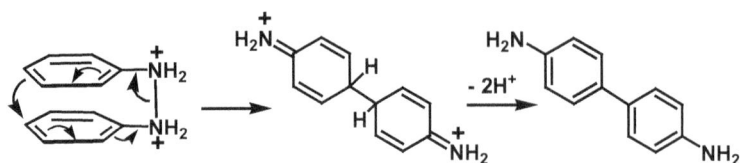

Scheme 2.20.   Mechanism of the benzidine rearrangement.

References

1. J.D. Roberts, H.E. Simmons Jr., L.A. Vaughan and C.W. Carlsmith, Rearrangement in the reaction of chlorobenzene-1-^{14}C with potassium amide. *J. Am. Chem. Soc.* **1953**, *75*, 3290–3291.
2. T.L. Gilchrist and C.W. Rees, Chapter 8: Reactions of arynes. In *Carbenes, Nitrenes and Arynes*; Thomas Nelson & Sons: London, UK, 1969; pp. 103–117.
3. W.D. McPhee and E. Klingsberg, The reaction of some α-chloroketones with alkali. *J. Am. Chem. Soc.* **1944**, *66*, 1132–1136.
4. R.B. Loftfield, The alkaline rearrangement of α-haloketones. II. The mechanism of the Favorskii reaction. *J. Am. Chem. Soc.* **1951**, *73*, 4707–4714.
5. P.E. Eaton and T.W. Cole Jr., Cubane. *J. Am. Chem. Soc.* **1964**, *86*, 3157–3158.
6. E.D. Hughes, F. Juliusburger, S. Masterman, B. Topley and J. Weiss, Aliphatic substitution and the Walden inversion. Part I. *J. Chem. Soc.* **1935**, 1525–1529.
7. E.D. Hughes, F. Juliusburger, A.D. Scott, B. Topley and J. Weiss, Aliphatic substitution and the Walden inversion. Part II. *J. Chem. Soc.* **1936**, 1173–1175.
8. W.A. Cowdrey, E.D. Hughes, T.P. Nevell and C.L. Wilson, Aliphatic substitution and the Walden inversion. Part III. Comparison, using radioactive bromine of the rules of inversion and substitution in the reaction of bromide ions with α-bromopropionic acid. *J. Chem. Soc.* **1938**, 209–211.
9. L. Tenud, S. Farooq, J. Seibl and A. Eschenmoser, Endocyclic S_N-reactions at saturated carbon. *Helv. Chim. Acta* **1970**, *53*, 2059–2068.
10. J. March, Chapter 10: Aliphatic nucleophilic substitution. In *Advanced Organic Chemistry, Reactions, Mechanism and Structure*, 4th ed.; John Wiley & Sons: New York, NY, USA, 1992; pp. 298–305.
11. R. Fittig, Some transformations of the acetone of acetic acid. *Liebigs Ann.* **1859**, *110*, 23–45.
12. A. Butlerow, About trimethylacetic acid. *Liebigs Ann.* **1873**, *170*, 151–162.
13. J. March, Chapter 18: Rearrangements. In *Advanced Organic Chemistry, Reactions, Mechanism and Structure*, 4th ed.; John Wiley & Sons: New York, NY, USA, 1992; pp. 1058–1061.

14. W.E. Bachmann and H.R. Sternberger, The pinacol-pinacolone rearrangement. V. The rearrangement of unsymmetrical aromatic pinacols. *J. Am. Chem. Soc.* **1934**, *56*, 170–173.

15. N. Weliky and E.S. Gould, Studies in the ferrocene series. I. Some reactions of compounds related to monobenzoylferrocene. *J. Am. Chem. Soc.* **1957**, *79*, 2742–2746.

16. P.L. Pauson and W.E. Watts, Ferrocene derivatives. Part XII. Di-and tri-ferrocenylmethane derivatives. *J. Chem. Soc.* **1962**, 3880–3886.

17. L.R. Moffett Jr., Pinacol and pinacolone derivatives of some acylferrocenes. *J. Org. Chem.* **1964**, *29*, 3726–3727.

18. M.D. Rausch and D.L. Adams, The Clemmensen reduction of benzoylferrocene. *J. Org. Chem.* **1967**, *32*, 4144–4145.

19. S.I. Goldberg and W.D. Bailey, Evidence for appreciable phenyl migration in the rearrangements of *threo* and *erythro*-1,2-diferroceny-1,2-diphenylethane-1,2-diols. *Chem. Commun.* **1969**, 1059–1060.

20. M. Görmen, P. Pigeon, E.A. Hillard, A. Vessières, M. Huché. M.-A. Richard, M.J. McGlinchey, S. Top and G. Jaouen, Synthesis and antiproliferative effects of [3]ferrocenophane transposition products and pinacols obtained from McMurry cross-coupling reactions. *Organometallics* **2012**, *31*, 5856–5866.

21. M.J. McGlinchey, Ferrocenyl migrations and molecular rearrangements: the significance of electronic charge delocalisation. *Inorganics* **2020**, *8*, 68, and references therein.

22. W. Kirmse, 100 years of the Wolff rearrangement. *Eur. J. Org. Chem.* **2002**, 2193–2256.

23. J. Fenwick, G. Frater, K. Ogi and O.P. Strausz, Mechanism of the Wolff rearrangement. IV. The role of oxirene in the photolysis of α-diazo ketones and ketenes. *J. Am. Chem. Soc.* **1973**, *95*, 124–132.

24. K.-P. Zeller, H. Meier, H. Kolshorn and E. Müller, On the mechanism of the Wolff rearrangement. *Chem. Ber.* **1972**, *105*, 1875–1886.

25. S. Winstein and H.J. Lucas, Retention of configuration in the reaction of the 3-bromo-2-butanols with hydrogen bromide. *J. Am. Chem. Soc.* **1939**, *61*, 1576–1581.

26. S. Winstein and H.J. Lucas, The loss of optical activity in the reaction of the optically active *erythro*- and *threo*-3-bromo-2-butanols with hydrobromic acid. *J. Am. Chem. Soc.* **1939**, *61*, 2845–2848.

27. D.J. Cram, Studies in stereochemistry, I. The specific Wagner-Meerwein rearrangement of the isomers of 3-phenyl-2-butanol. *J. Am. Chem. Soc.* **1949**, *71*, 3863–3870.

28. E. Beckmann, Learning about the isonitrosocompounds. *Ber. Dtsch. Chem. Ges.* **1886**, *19*, 988–993.

29. L.W. Jones and E.S. Wallis, The Beckmann rearrangement involving optically active radicals. *J. Am. Chem. Soc.* **1926**, *48*, 169–181.

30. K. Schlögl and H. Mechtler, On the rearrangement of acylferrocene oximes during reduction with lithium alanate-aluminium chloride. *Monatsh.* **1966**, *97*, 150–167.
31. Th. Curtius, Hydrazide und Azide organischer Säuren. *J. Prakt. Chem.* **1894**, *50*, 275–294.
32. J. Kenyon and D.P. Young, Retention of asymmetry during the Curtius and the Beckmann change. *J. Chem. Soc.* **1941**, 263–267.
33. R.B. Woodward, T. Fukunaga and R.C. Kelly, Triquinacene. *J. Am. Chem. Soc.* **1964**, *86*, 3162–3164.
34. K. Shimada, S. Orii, M. Tanaka and T. Nambara, New ferrocene reagents for derivatization of alcohols in high-performance liquid chromatography with electrochemical detection. *J. Chromatogr.* **1986**, *352*, 329–335.
35. S. Goggins, E.A. Apsey, M.F. Mahon and C.G. Frost, Ratiometric electrochemical detection of hydrogen peroxide and glucose. *Org. Biomol. Chem.* **2017**, *15*, 2458–2466.
36. H.J. Shine, Reflections on the ϖ theory of benzidine rearrangements. *J. Phys. Org. Chem.* **1989**, *2*, 491–506.
37. H.J. Shine, H. Zmuda, K.H. Park, H. Kwart, A.G. Horgan and M. Brechbiel. Benzidine rearrangements. 16. The use of heavy-atom kinetic isotope effects in solving the mechanism of the acid-catalyzed rearrangement of hydrazobenzene. The concerted pathway to benzidine and the nonconcerted pathway to diphenylene. *J. Am. Chem. Soc.* **1982**, *104*, 2501–2509.

Chapter 3

Experimental Validation of the Conservation of Orbital Symmetry

"Wherever a symmetry of nature occurs, there is a conservation law attached to it, and vice versa."

— *Emmi Noether*

The Woodward-Hoffmann rules are, obviously, symmetry-based, but many of their rationalisations were prompted by experimental observations, and their predictions verified by judiciously engineered symmetry breaking in the molecules studied. As has been well documented, in one step of Woodward's synthetic route to vitamin B_{12} there arose the need to explain why the stereochemical outcome of a particular ring closure depended on whether the reaction was initiated thermally or photochemically.[1] This was then reduced to the more fundamental question concerning the interconversion of butadienes and cyclobutenes, and also of hexatrienes and cyclohexadienes in concerted processes. In their masterful explication of the underlying principles of the conservation of orbital symmetry,[2] Woodward and Hoffmann provided a multitude of beautiful examples of its application, and we here discuss selected cases in which molecular symmetry breaking subsequently played a subtle but particularly important role.

3.1. Electrocyclisations

Focussing initially on electrocyclic ring openings and closures, molecular orbital calculations at the Extended Hückel (EHMO) level indicated that the cyclobutene-butadiene interconversion proceeds in a conrotatory manner under thermal conditions, but in a disrotatory fashion when in a photochemically excited state. In contrast, the cyclohexadiene-hexatriene interconversion proceeds disrotatorily in the ground state (i.e., thermally), but conrotatorily when photolysed.[3] Of course, the EHMO calculations were carried out on the parent hydrocarbons, but experimental verification required that these reactions be monitored on molecules of lower symmetry.

Fortunately, appropriate data were already available to Woodward and Hoffmann. Typically, the products arising from the thermolysis of *cis*, **1**, and *trans*, **2**, 1,2,3,4-tetramethylcyclobutene are depicted in Scheme 3.1, and evidently arise by conrotatory ring opening.[4] These intramolecular rotations can occur in either the clockwise or anticlockwise direction, but in the *cis* case the *cis,trans*-dienes that result are identical. However, starting from the *trans* cyclobutene, **2**, isomeric *cis,cis* and *trans,trans* dienes could arise; in the event, only the *trans,trans* product, **3**, is observed, presumably to avoid steric problems in the other isomer. In this early work by Criegee and Noll,[4] the isomers were characterised on the basis of their boiling points, refractive indices,

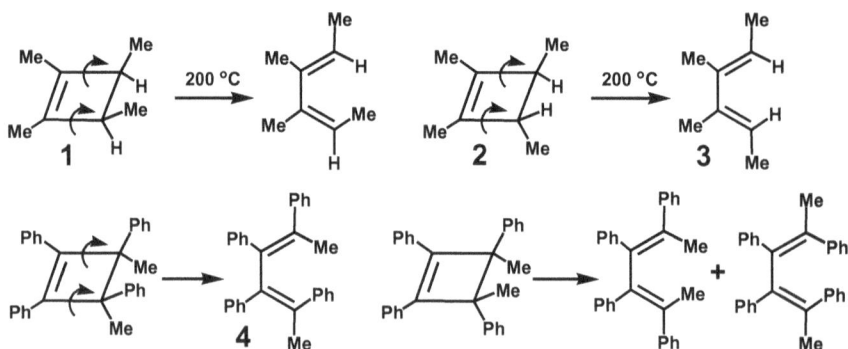

Scheme 3.1. Conrotatory thermal ring opening of *cis* and *trans* substituted cyclobutenes.

behaviour upon ozonolysis, and UV spectra. In a subsequent study, ring opening of *cis* and *trans* 3,4-dimethyl-1,2,3,4-tetraphenyl-cyclobutene was monitored directly by ^{1}H NMR spectroscopy; the *cis, trans* diene, **4**, exhibited individual methyl signals at 1.98 and 2.18 ppm, whereas in their *cis,cis* and *trans,trans* counterparts the two methyls were NMR equivalent, and resonated at 2.40 and 1.89 ppm, respectively.[5] Since that time, this phenomenon has been successfully verified in an enormous number of different situations and is now a standard synthetic procedure.

The electrocyclisation behaviour of the hexatriene-cyclohexadiene system was beautifully illustrated in complementary publications in the same issue of *Tetrahedron Letters* in 1965. In the more straight-forward case, Vogel reported that *trans,cis,trans*-2,4,6-octatriene, **5**, underwent the expected disrotatory thermal ring closure to form *cis*-5,6-dimethylcyclo-1,3-hexadiene, **6**.[6] The more instructive situa-tion, reported by Winstein,[7] arose when *cis,cis,cis*-1,4,7-cyclonona-triene, **7**, was treated with potassium t-butoxide and isomerised to form the conjugated system *cis,cis,cis*-1,3,5-cyclononatriene, **8**, which then suffered disrotatory ring closure to yield *cis*-bicyclo[4.3.0]nona-2,4-diene, **9**. This latter compound, **9**, was photolysed by Vogel and ring-opened in conrotatory fashion to furnish the *cis,cis,trans*-1,3,5-cy-clononatriene, **10**, that subsequently ring-closed thermally to form *trans*-bicyclo[4.3.0]nona-2,4-diene, **11**, as shown in Scheme 3.2. The overall effect is to bring about transformation of the

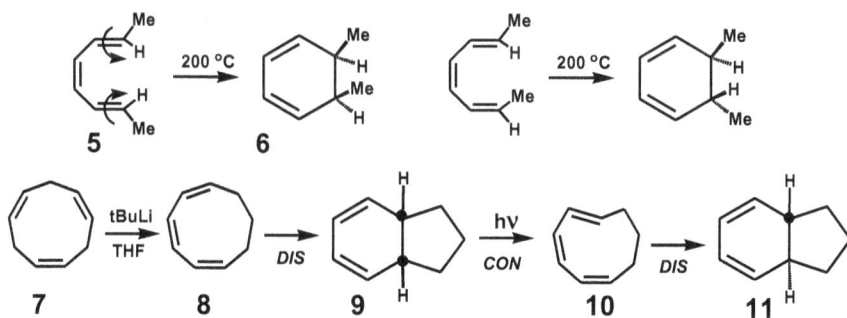

Scheme 3.2. Ring closings and openings in 1,3,5-hexatrienes and 1,3,5-cyclononatrienes.

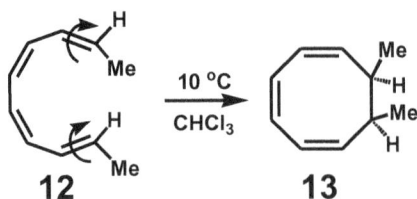

Scheme 3.3.   Conrotatory thermal ring-closing of a cyclooctatetraene.

mirror-symmetric (C_s) dihydroindane, **9**, via an intermediate cyclon-onatriene, **10**, to generate the C_2-symmetric product **11**. The products were all characterised spectroscopically (NMR, IR, UV).

Continuing in this vein, thermolysis of *cis,cis,cis,trans*-2,4,6,8-decatetraene, **12**, (a terminally substituted dimethyloctatetraene) brought about conrotatory ring closing to form *cis*-7,8-dimethylcyclo-1,3,5-octatriene, **13** (Scheme 3.3); as predicted, photolysis leads to the analogous *trans* isomer.[8] We return to discuss these reactions in more detail in Section 3.2.

Another particularly interesting prediction concerned the behaviour of a bicyclic system bearing a cyclopropyl ring that could open to generate an allylic cation. The favoured EHMO-calculated reaction indicated that the disrotatory ring opening should proceed such that the substituents on the same face of the three-membered ring as the leaving group rotate towards each other, thereby eliminating the *endo* substituent.[3]

Since it was established that 7,7-dichlorobicyclo[4.1.0]heptane, **14**, underwent solvolysis to form an acetate with elimination of chloride, it was necessary to break the local symmetry to determine which chloride, *endo* or *exo*, was eliminated. This was accomplished by addition of chlorocarbene to cyclohexene, thereby replacing one of the chlorides in **14** by hydrogen. The resulting epimers were separated by vapour phase chromatography and their identities determined by comparison of the coupling constants between H(7) and the other cyclopropane ring protons (7.5 and 3.3 Hz for the *endo* and *exo* epimers, respectively). Gratifyingly, it was found that the *endo* isomer, **15**, undergoes ready solvolysis whereas its *exo* counterpart is inert.[9] This result also allowed the resolution of the

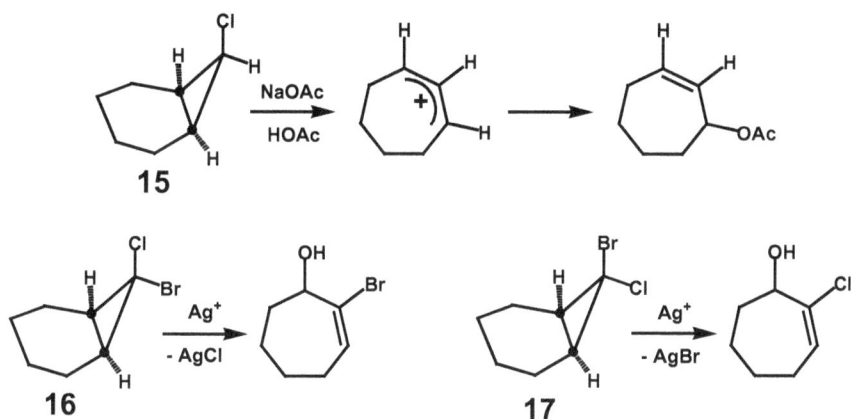

Scheme 3.4. Symmetry-controlled ring opening of cyclopropyl rings to allyl cations.

previously unknown assignments of the epimers **16** and **17**, reported by Skell and Sandler,[10] which gave different solvolysis products. In one case chloride was eliminated, while the other lost bromide; since the *endo* halide is the one displaced, the identity of each of the epimers is now evident and the observed products could now be rationalised, as depicted in Scheme 3.4.

3.2. Cycloadditions and Cycloreversions

Orbital symmetry analysis predicted that thermal (i.e., ground state) cycloadditions should occur in a *suprafacial-suprafacial* or *antarafacial-antarafacial* fashion for systems in which the reactants possess $4n+2$ π electrons.[11] The classic $[_\pi 4_s + _\pi 2_s]$ case is, of course, the Diels-Alder reaction, examples of which are multitudinous. In contrast, *suprafacial-antarafacial* cycloaddition is to be expected when the reactants present a total of $4n$ π electrons. Under photochemical conditions, these predictions would of course be reversed. This led to a search for previously unknown, or unrecognised, cycloadditions.[12] These included the $[_\pi 6_s + _\pi 4_s]$ reaction, as in **18**, from cyclopentadiene and tropone,[13] the $[_\pi 12_s + _\pi 2_s]$ cycloadduct, **19**, formed by reaction of a sesquifulvalene with tetracyanoethylene (a $4n+2$ π system),[14] and the contrasting $4n$ case whereby TCNE adds to a heptafulvalene in a

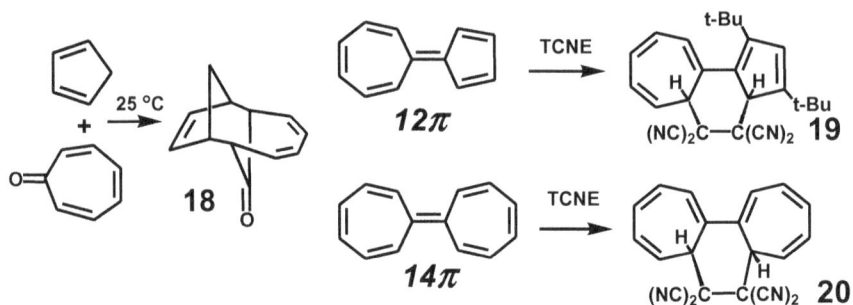

Scheme 3.5. Examples of [6+4], [12+2] and [14+2] cycloadditions.

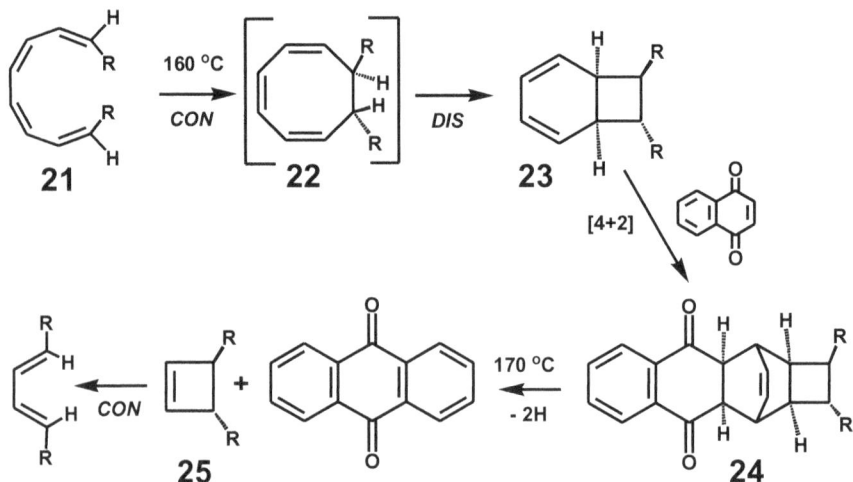

Scheme 3.6. A remarkable sequence of symmetry-allowed processes (R = methoxy).

$[_\pi 14_a + _\pi 2_s]$ fashion to yield **20** (Scheme 3.5).[15] In this latter case, the *trans* stereochemistry of the product was established unequivocally by X-ray crystallography.

A noteworthy series of reactions reported by Meister[16] is shown in Scheme 3.6 and provides a spectacular combination of ring closings and ring openings, together with cycloadditions and cycloreversions, each of which faithfully follows the phenomenon of orbital symmetry control.

The conrotatory ring closure of the 8π dimethoxytetraene, **21**, at 110–170°C, followed by the disrotatory closure of the resulting 6π cyclocta-1,3,5-triene, **22**, furnished dimethoxybicyclo[4.2.0]octa-2,4-diene, **23**. This molecule in turn underwent a $[_\pi 4_s +_\pi 2_s]$ Diels-Alder reaction with benzoquinone when heated at reflux in benzene, whereupon at 170°C the cycloadduct, **24**, lost two hydrogens and then suffered a retro-Diels-Alder to form anthraquinone and eliminate *trans*-3,4-dimethoxycyclobutene, **25**, that itself ring-opened conrotatorily to yield *trans,trans*-1,4-dimethoxybutadiene.[16]

3.3. Sigmatropic Shifts

3.3.1. *Suprafacial [1,5] and Antarafacial [1,7] Migrations*

The concerted migration of a σ-bonded atom or group across a π system is termed a sigmatropic shift. It can be designated as either *suprafacial* or *antarafacial* depending on whether the migrating moiety remains on the same face of the π system or is transferred onto the opposite face, respectively. The separation between the initial and final migration sites is incorporated into the nomenclature, as in suprafacial [1,5] or antarafacial [1,7] shifts.[17] The Cope rearrangement[18] can now be classified as a suprafacial [3,3] sigmatropic shift. Once again, symmetry breaking has been judiciously invoked to verify the predictions experimentally.

In a typical thermal (i.e., ground state) suprafacial [1,5] sigmatropic shift, stereospecific migration of a hydrogen atom between the termini of a labelled 1,5-pentadiene, **26** → **27**, with concomitant rearrangement of the double bonds has been observed.[19] However, in [1,7] antarafacial migrations the molecule must be capable of adopting a non-planar helical conformation[20] so as to allow the transfer from one face of the molecule to the other, as in the transformation of precalciferol into vitamin D_3.[21] A more readily visualisable example is found in the interconversion of the *cis,cis*-1,3,5-octatriene, **28**, and the *cis,cis*-2,4,6-octatriene, **29**, shown in Scheme 3.7.[22]

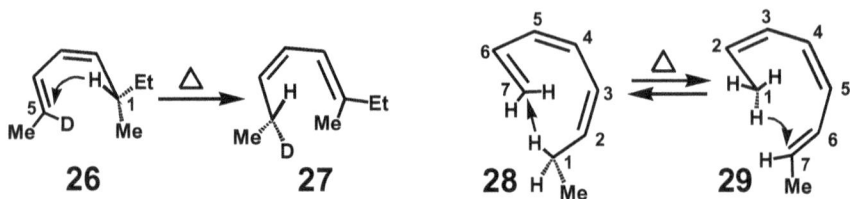

Scheme 3.7. Examples of suprafacial [1,5] and antarafacial [1,7] sigmatropic shifts.

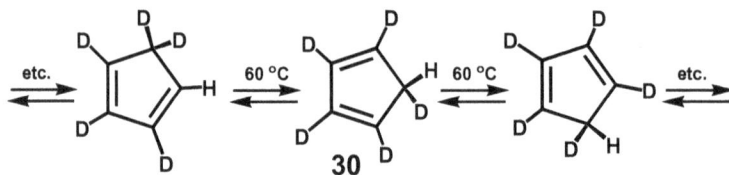

Scheme 3.8. Suprafacial [1,5] shift of hydrogen (or deuterium) across a cyclopentadiene ring.

3.3.2. [1,5] Shifts in Cyclopentadienes and Indenes

Following early work by Mironov,[23] hydrogen migration across an unsubstituted cyclopentadienyl ring proceeds with an activation energy barrier of 24 kcal mol^{-1}, as shown in an elegant experiment by Roth in which the inherent symmetry of the system was broken by studying the behaviour of 5*H*-pentadeuterocyclopentadiene, **30**, over the range 45–65°C. This process also scrambled the deuteriums such that the hydrogen gained access to all the vinylic positions (Scheme 3.8). In contrast, migrations of alkyl groups, such as t-butyl, only occur at much higher temperatures (>220°C) and the barrier is in excess of 40 kcal mol^{-1}.[24]

However, organometallic and organometalloidal moieties rearrange with relatively low migration barriers. Indeed, one such reaction holds an important historical position — the first case of a fully elucidated organometallic fluxional process. In pioneering work at Harvard in 1956,[25] T.S. Piper and Geoffrey Wilkinson (Nobel Prize 1973, for the discovery of ferrocene) prepared a molecule, **31**, of formula $(C_5H_5)_2Fe(CO)_2$, clearly defying the conventional 18-electron rule! Moreover, the ^1H NMR spectrum of **31** at room temperature was surprisingly simple and exhibited only two equal intensity singlets. Surprisingly, it took another 10 years before this observation was revisited, by a group at MIT who undertook a low-temperature ^{1}H NMR study of **31**. Upon cooling the sample, one of the singlets decoalesced to yield a 2:2:1 peak pattern, and it was recognised that while one cyclopentadienyl ring was π-complexed (now designated as η^5-C_5H_5) the other was σ-bonded to only one ring carbon at a time (η^1-C_5H_5), a result that was confirmed by X-ray crystallography. Eventually, it was established that the (η^5-C_5H_5) Fe(CO)$_2$ fragment was undergoing a series of rapid stepwise migrations between adjacent ring carbons with a barrier of ~10.7 kcal mol^{-1}.[26] At the time, it was described as a 1,2-shift, but was subsequently reformulated in orbital symmetry terms as a suprafacial [1,5] sigmatropic shift. Interestingly, similar migration behaviour was also found in trimethylsilylcyclopentadiene, **32**, as shown in Scheme 3.9, and also for analogous systems containing a wide range of other elements in Group 13 (B, Al, Ga, In), Group 14 (Ge, Sn, Pb) and Group 15 (P, As, Sb).[27]

The goal of extending this behaviour to its organometallic indenyl counterpart, (η^5-C_5H_5)Fe(CO)$_2$(η^1-C_9H_7) **33**, was frustrated by its failure to exhibit fluxionality at room temperature, and by loss of the carbon monoxide ligands to form benzoferrocene when heated too strongly.[28] It was suggested that the enhanced barrier to migration was the result of its reluctance to pass via an isoindene intermediate with the consequent loss of aromatic character. This followed the pattern found in the corresponding organic systems for which the

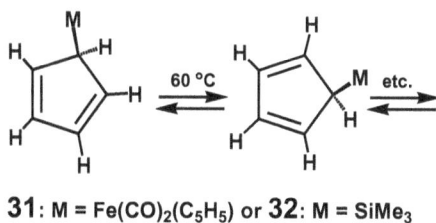

31: M = Fe(CO)$_2$(C$_5$H$_5$) or **32**: M = SiMe$_3$

Scheme 3.9. Suprafacial [1,5] metallotropic shifts across a cyclopentadienyl framework.

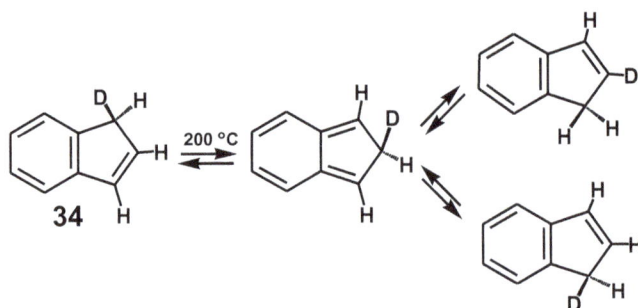

Scheme 3.10. Deuterium scrambling by [1,5] sigmatropic shifts via an isoindene intermediate.

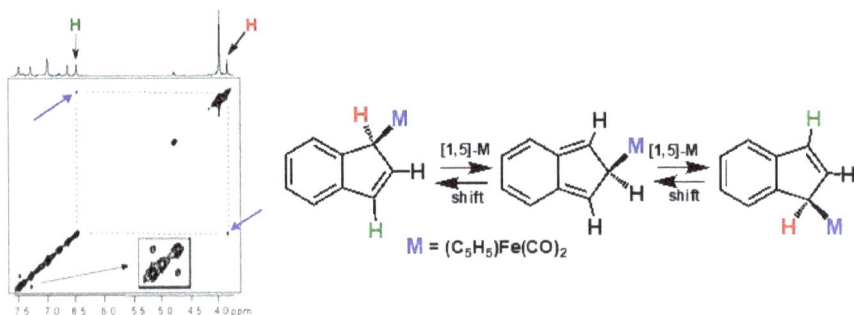

Figure 3.1. 500 MHz 2D-EXSY spectrum of **33** at 45°C with off-diagonal cross-peaks showing NMR equilibration of the indenyl protons H(1) and H(3).[29]

reported activation energies for hydrogen migrations in cyclopentadiene and indene were approximately 24 and 35 kcal mol^{-1}, respectively. In the latter case, it was shown that, starting from the monodeuteroindene, **34**, the label was scrambled over all the non-aromatic sites at elevated temperatures (Scheme 3.10).[24]

This difficulty was only overcome several decades later by taking advantage of the evidence provided by 2D-EXSY (exchange spectroscopy) NMR whereby, even at temperatures below that required for peak coalescence, exchange between sites can be monitored via the off-diagonal cross-peaks, as illustrated in Figure 3.1.

The migration barrier thus obtained was found to be ~24 kcal mol^{-1}; confirmation of the intermediacy of the isoindene was

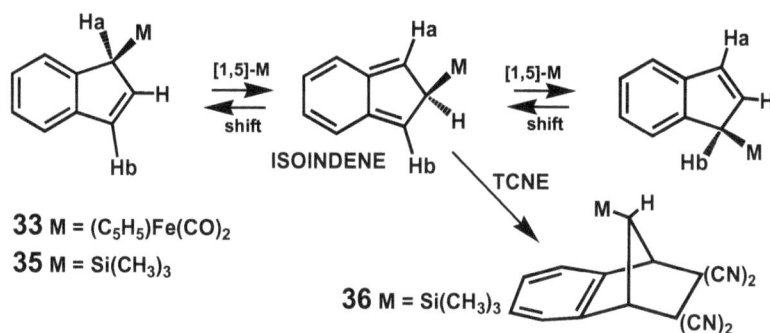

Scheme 3.11. Metallotropic [1,5] shifts of organo-iron and organo-silicon moieties, and Diels-Alder trapping of the isoindene intermediates.[28,29]

demonstrated through its trapping by tetracyanoethylene as its Diels-Alder cycloadduct that was unequivocally characterised by X-ray crystallography.[29] This approach has also been adopted for the case of trimethylsilylindene, **35**; once again, 2D-EXSY data revealed the [1,5] migration of the silyl substituent via an isoindene whose intermediacy was also confirmed by X-ray crystallographic characterisation of its Diels-Alder TCNE cycloadduct, **36** (Scheme 3.11).[30] In these cases, it is evident that, starting from an asymmetrical (C_1) indene, rearrangement via a mirror-symmetric (C_s) isoindene intermediate interconverts enantiomers, a point to which we shall return.

Moreover, in a series of beautifully executed experiments, whereby the incorporation of diastereotopic methyls that maintain their non-equivalence, as in **37**, provided a method to monitor any change in stereochemistry, Stobart demonstrated unambiguously that these migrations proceed with retention of configuration of the migrating group (Scheme 3.12).[31]

It was also demonstrated that the incorporation of additional benzo rings, as in Scheme 3.13, to extend the aromatic framework from 6π to 10π to a 14π electron system markedly lowers the barrier to migration via an isoindene intermediate.[32,33] Indeed, the isoindene, **38**, generated by thermolysis of dibenz[*e,g*]indene, **39**, is sufficiently long-lived that it undergoes Diels-Alder cycloaddition with its own precursor (Scheme 3.14).[34]

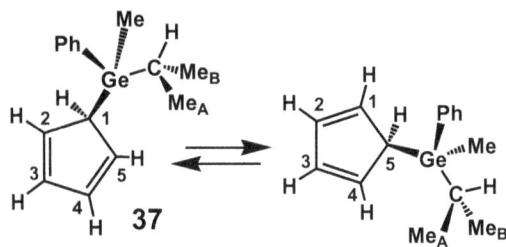

Scheme 3.12. Retention of stereochemistry during a suprafacial [1,5] sigmatropic shift.

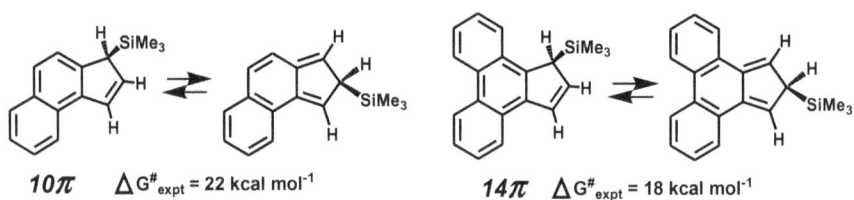

10π $\Delta G^{\#}_{expt} = 22$ kcal mol^{-1} 14π $\Delta G^{\#}_{expt} = 18$ kcal mol^{-1}

Scheme 3.13. Incorporation of additional benzo rings lowers the barrier for a [1,5] shift.

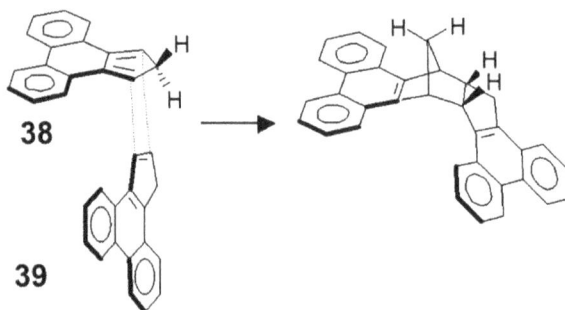

Scheme 3.14. [4+2] Cycloaddition of dibenz[e,g]indene with its own isoindene.

This project has been extended to encompass molecules bearing multiple η^1-indenyl substituents as in bis(indenyl)dimethylsilane, **40**,[35] tris(indenyl)methylsilane, **41**,[30] and even tetrakis(indenyl)-silane, -germane and -stannane.[36,37] We can see from Scheme 3.15 that in bis(indenyl)dimethylsilane consecutive [1,5] shifts, whereby a silicon migrates from C(1) via C(2) to C(3), interconvert diastereomers.

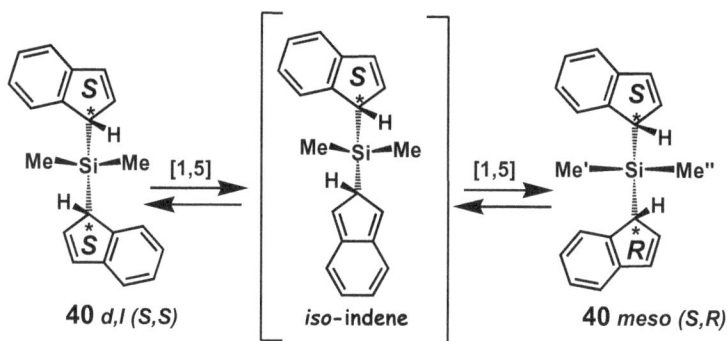

Scheme 3.15. Interconversion of *d,l* and *meso* diastereomers of bis(η^1-indenyl) dimethylsilane.

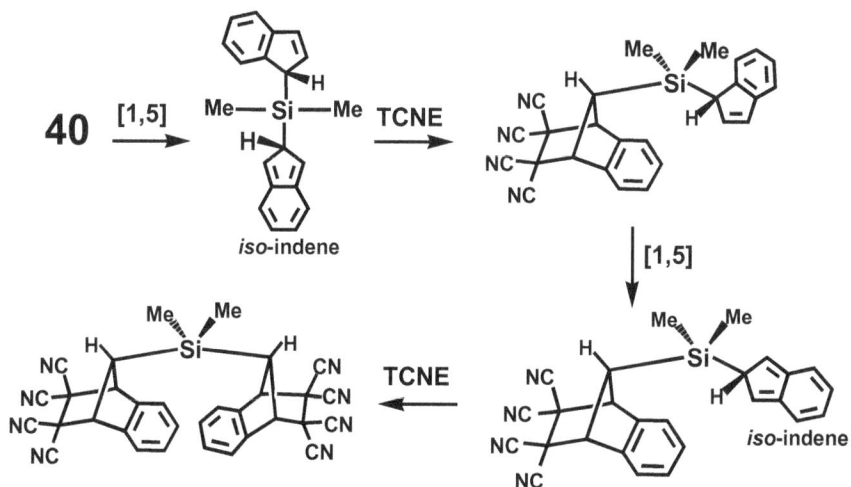

Scheme 3.16. Sequential Diels-Alder trapping by TCNE of the isoindenes derived from bis(η^1-indenyl)dimethylsilane, **40**.

One should note that the methyl groups in the C_2 isomers (*S,S*) and (*R,R*) are symmetry equivalent, whereas in the *meso* compound (*R,S*) they are rendered non-equivalent since they lie in the molecular mirror plane and appear as separate NMR singlets. Monitoring exchange between these methyl environments allowed evaluation of the migration barrier as ~23 kcal mol^{-1} and, gratifyingly, the system gave rise to a double Diels-Alder adduct with TCNE (Scheme 3.16).[35]

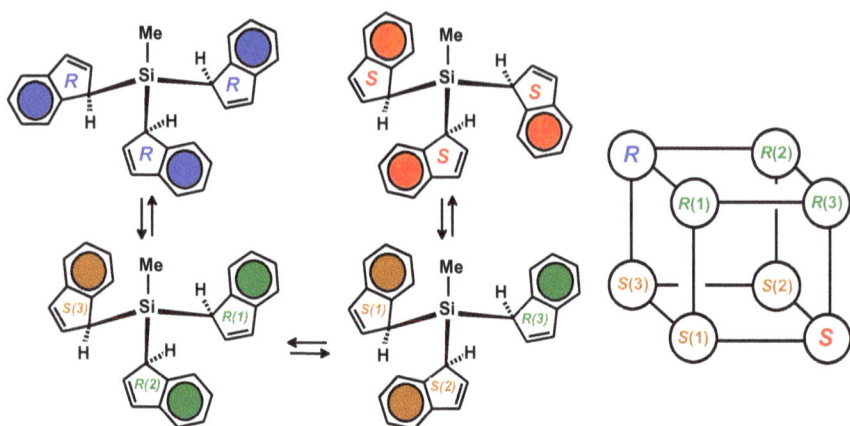

Scheme 3.17. Interconversions of the eight indenyl ring environments in **41**. The configurational inversion of a single indenyl ring requires two [1,5] suprafacial sigmatropic shifts.

Tris(η^1-indenyl)methylsilane, **41**, has been comprehensively studied since it offers a particularly fascinating stereochemical situation in that it exists as two pairs of enantiomers whose interconversions can be mapped onto a cube. Thus, starting from the C_3-symmetric *R,R,R* isomer, migration of silicon across a single indenyl ring yields the *R,R,S* diastereomer in which all three indenyls are NMR inequivalent. Two further sets of silicon shifts produce the *S,S,S* configuration, enantiomeric to the starting structure, and this complete exchange pathway can be represented by migrations along the edges of a cube (Scheme 3.17).[30]

A section of the 500 MHz ^{1}H 2D-EXSY spectrum of **41** is shown as Figure 3.2. It not only exhibits off-diagonal cross-peaks connecting those sites undergoing direct exchange, but also reveals two-step processes, such as **R1** to **R2** or **S2** to **R3** in Scheme 3.17, as less intense cross-peaks.

As with the earlier cases, tris(η^1-indenyl)alkylsilanes undergo successive [4+2] cycloadditions with TCNE, ultimately yielding triple Diels-Alder adducts, **42**, that have been characterised by X-ray crystallography (Scheme 3.18).[38,39] The whole area of syntheses and molecular dynamics of η^1-indenyl derivatives of main group elements and transition metals has been comprehensively reviewed.[40]

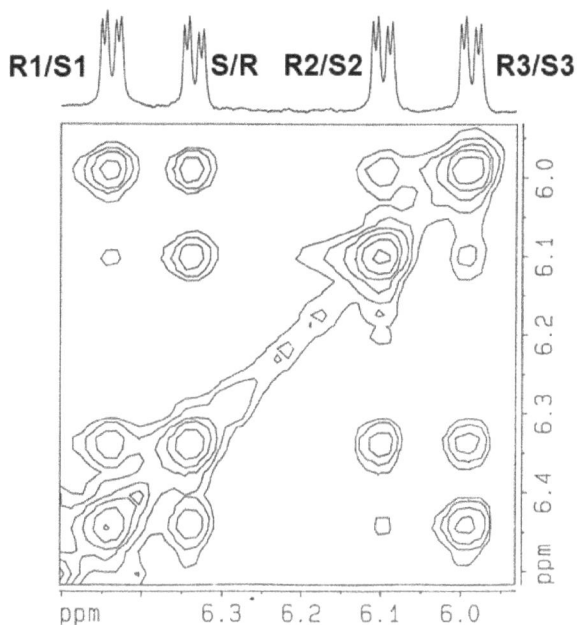

Figure 3.2. Section of the 2D-EXSY spectrum of **41** of the indenyl H(2) resonances: cross-peaks for one-step exchanges are large, whereas those requiring two steps are less intense.[30]

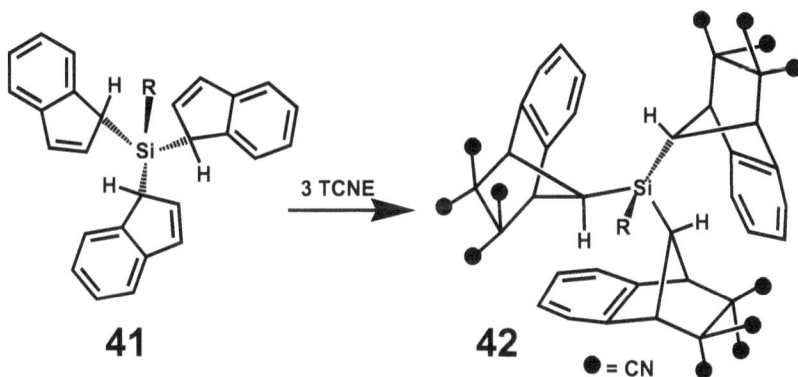

Scheme 3.18. Formation of a triple Diels-Alder cycloadduct with tetracyanoethylene.

3.3.3. *[1,5] and [1,7] Sigmatropic Shifts in Cycloheptatrienes*

As noted above, from orbital symmetry considerations one can predict that suprafacial [1,5] and [1,7] migrations should proceed exclusively under thermal and photochemical conditions, respectively. This has been beautifully demonstrated in the case of 1,4-bis(cycloheptrienyl)-benzene, **43**, as depicted in Scheme 3.19. When heated at 225°C, the product formed, **44**, is that expected for two [1,5] shifts, whereas photolysis furnished the corresponding [1,7] isomer, **45**.[41]

It is noteworthy that the boat-shaped conformation of cycloheptatrienes is particularly favourable to facilitate suprafacial [1,5] shifts. This has been clearly exemplified in a series of reactions whereby the cycloheptatriene has been formed by cheletropic loss of carbon monoxide from the Diels-Alder adducts of cyclopentadienones and cyclopropenes.[42] Thus, when allowed to react at room temperature, tetraphenylcyclopentadienone and 1,2,3-triphenylcyclopropene yield initially the [4+2] cycloadduct, **46**, that possesses an *endo* hydrogen in the three-membered ring. Upon heating at 145°C in refluxing xylene, loss of CO led to ring opening and formation of heptaphenylcycloheptatriene, **47**, in which this hydrogen is now oriented in an *exo* position. For both **46** and **47** the structures were established unambiguously by X-ray crystallography.[43] Evidently, the molecule has undergone a ring flip (Scheme 3.20), presumably to relieve steric strain among the bulky phenyl groups.

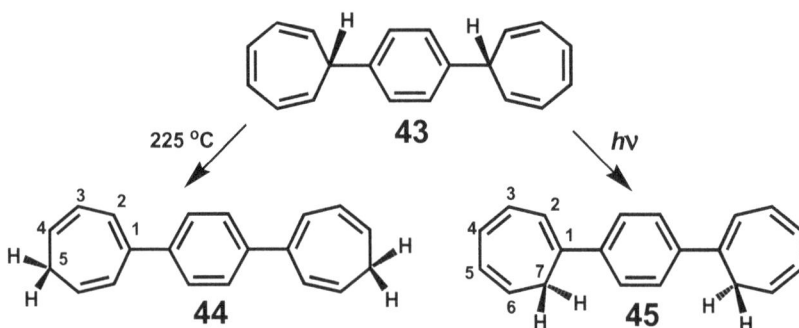

Scheme 3.19. Thermal [1,5] and photochemical [1,7] suprafacial sigmatropic shifts.

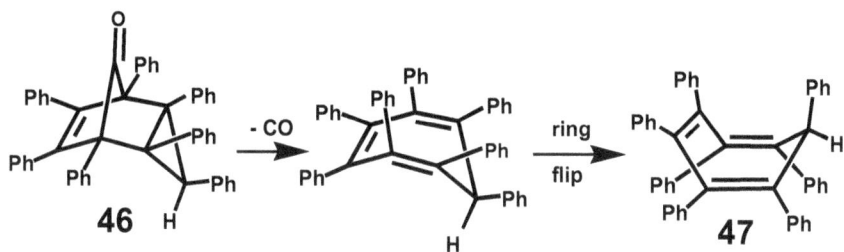

Scheme 3.20. Cheletropic loss of CO from the Diels-Alder adduct, **46**, leads to formation of heptaphenylcycloheptatriene, **47**.

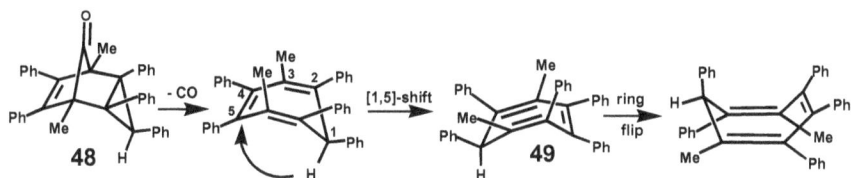

Scheme 3.21. Suprafacial [1,5] sigmatropic shifts in dimethylpentaphenylcyclohep-tatriene prior to a ring flip.

However, when the structure of the initially formed cycloadduct, **48**, was modified by the incorporation of methyl substituents in the cyclopentadienone precursor, the dimethylpentaphenylcycloheptatriene so formed had clearly undergone a series of rapid suprafacial [1,5] hydrogen shifts *while in its initially generated conformation*, **49**, before the ring flip stopped any further migrations (Scheme 3.21).[43]

Likewise, as shown in Scheme 3.22, the analogous cyclohepta-triene generated by reaction of 3,4-di-*p*-tolyl-2,5-diphenylcyclopenta-dienone and 1,2,3-triphenylcyclopropene was formed as four isomers in the statistically expected 1:2:2:2 ratio.[43]

In another clear example of such a rearrangement, the reaction of 3-ferrocenyl-2,4,5-triphenylcyclopentadienone, **50**, and 1,2,3-triphe-nylcyclopropene yielded only 1-ferrocenyl-2,3,4,5,6,7-hexaphenyl-cycloheptatriene, **51**, which was also characterised by X-ray crystallography (Scheme 3.23).[44]

We note in passing that symmetry-allowed [1,2] shifts in carboca-tions are very common, and there are examples whereby such a

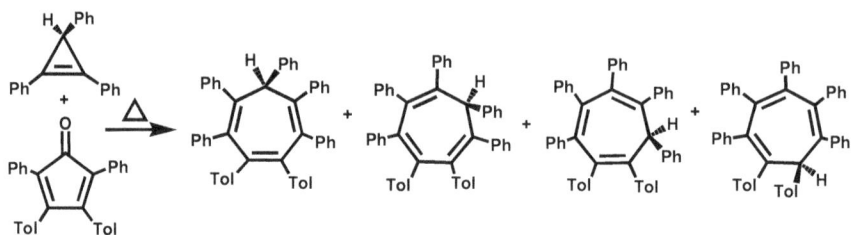

Scheme 3.22. Formation of four isomers of di-*p*-tolyl-pentaphenylcycloheptatriene.

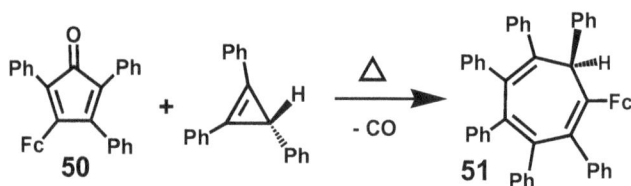

Scheme 3.23. Suprafacial [1,5] sigmatropic shifts in ferrocenyl-hexaphenylcycloheptatriene.

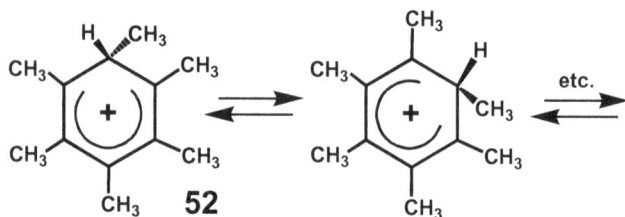

Scheme 3.24. [1,2] or [1,6] migration of a proton around a hexamethylbenzene ring.

migration could also be regarded as a [1,6] shift. For example, proto-nation of hexamethylbenzene yields a benzenonium cation, **52**, in which the single hydrogen migrates readily round the ring (Scheme 3.24). This gives rise to a spectacular ^{1}H NMR spectrum in which the migrating hydrogen couples equivalently to the eighteen hydrogens in the six methyl groups resulting in a 19-line multiplet with intensi-ties appropriate for the coefficients of the binomial expansion.[45]

3.3.4. *[3,3] and [5,5] Sigmatropic Shifts*

The classic suprafacial [3,3] shift is, of course, the Cope rearrangement. This reaction was originally reported in 1940 by Cope and Hardy,[18] then at Bryn Mawr College in Pennsylvania. The former subsequently held several important academic positions (at Columbia and MIT), and the Arthur C. Cope Award in his honour is now given out annually by the American Chemical Society for outstanding work in organic chemistry. Elizabeth Hardy went on to a distinguished industrial career with American Cyanamid.

As part of their studies on vinylated molecules, what their nomenclature described as ethyl (1-methylpropenyl)allylcyanoacetate, **53**, was obtained from the base-promoted reaction of ethyl (1-methylpropenylidene)cyanoacetate with an allyl halide. However, when heated at 150–160°C, **53** isomerised to form ethyl (1,2-dimethyl-4-pentenylidene)-cyanoacetate, **54**, as shown in Scheme 3.25.[18] With remarkable perspicacity, the authors recognised that what they termed an "intramolecular α,γ allyl shift" had occurred, and pointed out the close analogy to the rearrangements of vinyl allyl ether, **55**, and allyl phenyl ether, **56**, into allyl acetaldehyde and *o*-allyl phenol, respectively (Scheme 3.26).

These [3,3] sigmatropic shifts have been so thoroughly investigated and so widely extended that we have chosen to focus only on those aspects in which changes in symmetry have been an important

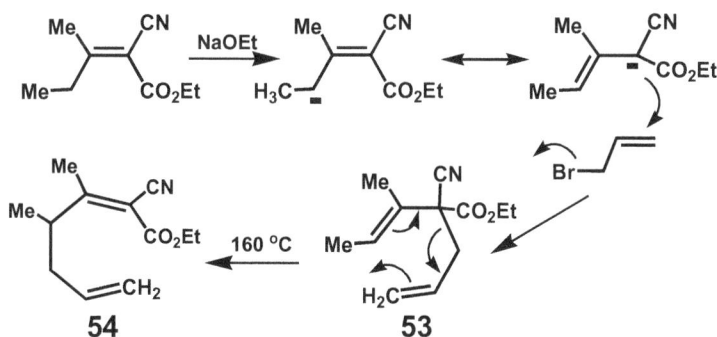

Scheme 3.25.   The original rearrangement discovered by Cope and Hardy.

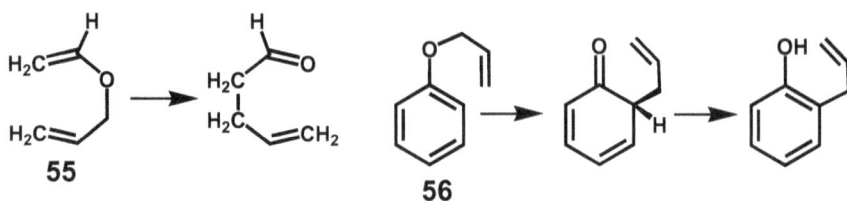

55

56

Scheme 3.26.   Cope rearrangement of vinyl allyl ether and phenyl allyl ether.

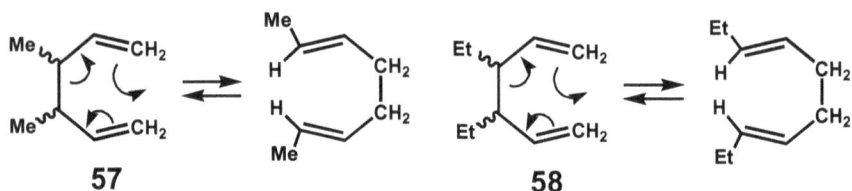

57   **58**

Scheme 3.27.   Cope rearrangements of 1,5-hexadienes.

factor. In particular, we consider the compelling evidence for the intramolecular character of the process, and also those reactions in which a cyclopropyl ring plays a prominent role.

In its simplest form, the Cope rearrangement is found as the thermally allowed reversible interconversion of 1,5-hexadienes. It is known that in suitably labelled systems, the only products seen are those arising from concerted intramolecular rearrangement, as in Scheme 3.27.[46] If the reaction were to proceed via two independent allyl radicals, then an isomeric product could be formed. Moreover, in "mixed Cope reactions" whereby two different 1,5-hexadienes, such as 3,4-dimethyl-1,5-hexadiene, **57**, and 3,4-diethyl-1,5-hexadiene, **58**, are heated together, "crossed products" such as 3-methyl-4-ethyl-1,5-hexadiene are not formed. We note, however, that in special circumstances whereby substituents such as phenyls can resonance-stabilise intermediate radicals, then mixed products can be detected.

Although [3,3] sigmatropic shifts generally favour a chair-like transition state, those involving a boat-like geometry are also well represented. Typically, attempts to isolate *cis*-1,2-divinylcyclopropane, **59**, even at –40°C, yielded instead 1,4-cycloheptadiene, **60**; in contrast, the corresponding *trans* isomer only rearranged at 200°C via a non-concerted diradical intermediate.[47,48] Similar behaviour was also seen in bicyclo[5.1.0]octa-2,5-diene, **61**,[49] and also in semibull-valene, **62**,[50] both of which underwent fast degenerate rearrangement even at low temperatures (Scheme 3.28).

This phenomenon reaches a climax in bullvalene, **63**, brilliantly prepared by thermolysis of cyclooctatetraene leading to **64**, one of the several dimers formed; UV irradiation resulted in elimination of benzene and formation of bullvalene (Scheme 3.29).[51]

In bullvalene the three all-*cis* vinyl substituents on the cyclopropyl ring are in turn connected to a single methine moiety. The resulting molecule has C_{3v} symmetry and undergoes Cope rearrangement (Scheme 3.30) such that any of the 10 CH units can occupy any position, giving rise to $10!/3 = 1,209,600$ identical isomers. It exhibits

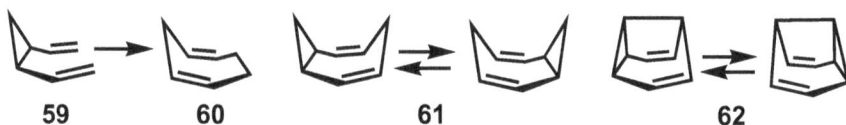

Scheme 3.28.   Cope rearrangements of vinyl cyclopropanes.

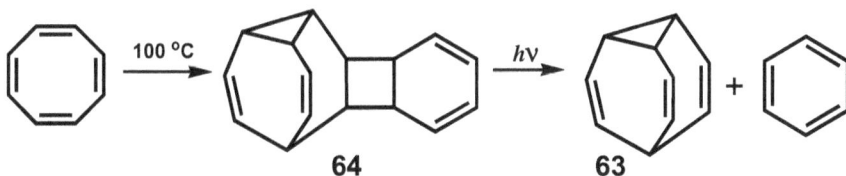

Scheme 3.29.   Schröder's remarkably efficient synthetic route to bullvalene.

Scheme 3.30.   Multiple Cope rearrangements equilibrate all 10 positions in bullvalene, **63**.

^{1}H and ^{13}C NMR singlets at 120°C, and only at –60°C is the process sufficiently slowed so as to reveal its three-fold symmetry with ^{13}C resonances in the ratio 1:3:3:3;[52,53] this structure has also been established in the solid state by X-ray crystallography.[54]

As noted previously, the benzidine rearrangement can be regarded as a [5s,5s] sigmatropic shift involving cleavage of the N–N linkage and formation of a bond between the C(4) and C(4′) positions (Scheme 3.31). Its intramolecular character is well documented in crossover experiments such as the case in which the ^{15}N,^{15}N-labelled precursor does not scramble with its normal ^{14}N,^{14}N counterpart.

In the search for other [5,5] sigmatropic shifts, Hafner has reported that the 5,5a,10,10a-tetrahydroheptalene system exhibits a molecular rearrangement consistent with such a scenario. In this case, the symmetry of the parent structure was broken by the introduction of appropriately positioned substituents.[55] This situation has been analysed by Houk who discussed initially the more readily visualised (*Z*,*Z*)-1,3,7,9-decatetraene, **65**, which is sufficiently flexible to allow both [5s,5s] and [5a,5a] shift stereochemistries.[56] As depicted in Scheme 3.32, the unlabelled product is identical to its precursor.

Returning now to Hafner's system, **66**, at 140°C it exchanges with its isomeric form, **67**, as revealed by the repositioning of the

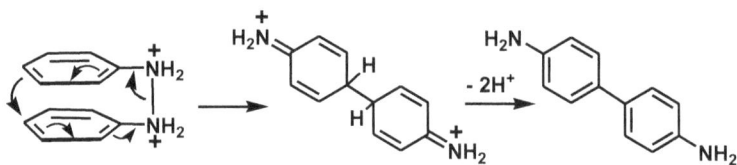

Scheme 3.31. Benzidine rearrangement viewed as [5,5] sigmatropic shift.

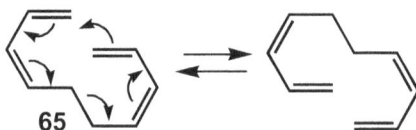

Scheme 3.32. A [5,5] sigmatropic shift in (*Z*,*Z*)-1,3,7,9-decatetraene.

substituents (Scheme 3.33). Although at first glance this appears to involve interactions between distant atoms, in fact the molecule can adopt a "cage-like" geometry in which bond scission (of the C(5)–C(6) linkage) and bond formation (between C(1) and C(10)) are perfectly viable (Scheme 3.34).[55]

In Houk's analysis, based on calculations at the DFT level, he concluded that a stepwise diradical mechanism is favoured.[56] More recently, Greer and Hoffmann have offered the suggestion that "*it is useful to regard the transition structure of the Cope rearrangement actually as a continuum of concerted, diradical and coupled allyl structures changing in response to the substituent pattern at C(2) and C(5), as well as at C(1), C(3), C(4) and C(6) . Thus, the Cope rearrangement has been called chameleonic.*"[57]

A particularly interesting prediction, also based on DFT calculations, concerning the possibility of [3,3], [5,5] and [7,7] sigmatropic shifts in a single molecular system has been offered by Kertesz.[58,59] X-ray data for 2,5,8-tri-*tert*-butyl-1,3-diazaphenalenyl (a substituted analogue of phenalenyl) revealed that the radicals exist as a σ-bonded

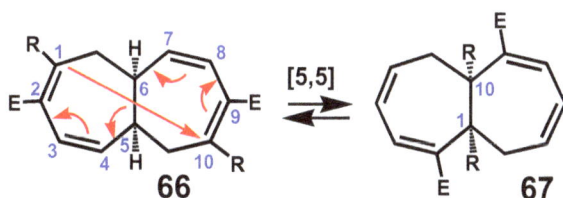

Scheme 3.33. Thermally allowed [5,5] sigmatropic shifts in dimethyl 5,5a,10,10a-tetrahydroheptalene-dicarboxylates.

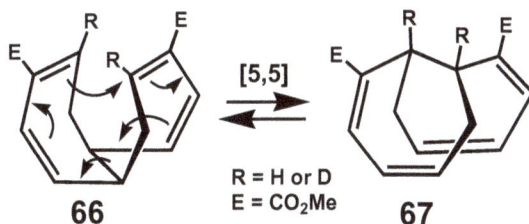

Scheme 3.34. Interconversion of **66** and **67** in a "cage-like" geometry.

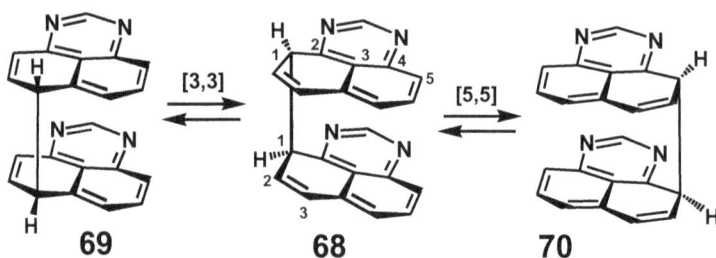

Scheme 3.35. Thermally allowed [3,3] and [5,5] sigmatropic shifts in the *syn* isomer **68**. (For clarity, the separation between the two halves of the dimer has been exaggerated.)

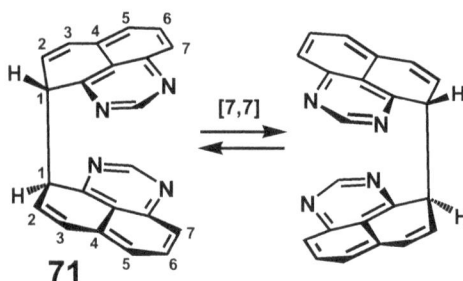

Scheme 3.36. A thermally allowed [7,7] sigmatropic shift in the *anti* isomer **71**.

dimer in the solid state.[60] It is proposed that the *syn* isomer, **68**, as seen in the solid state structure of the *tert*-butyl derivative, could undergo both non-degenerate [3,3] (**68 → 69**) and degenerate [5,5] (**68 → 70**) sigmatropic shifts, as in Scheme 3.35, passing via a π-dimer intermediate. In contrast, one could visualise the formation of a C_2-symmetric *anti* isomer, **71**, that could racemise via a [7,7] sigmatropic shift (Scheme 3.36).

These particular instances, selected from a very large number of reported examples, show the enduring influence of the principle of conservation of orbital symmetry on the determination of the reaction mechanisms and molecular rearrangements. Not only did Woodward and Hoffmann provide rationalisations of a plethora of previously unexplained experimental observations, but they also made many bold predictions that have since been validated. Indeed, reexamination of the original proposals by using calculations of increasing sophistication have enhanced our understanding of these processes. The importance of symmetry breaking in the experimental verification of the Woodward-Hoffmann rules is evident, and will undoubtedly continue to play a crucial role.

References

1. R.B. Woodward, Aromaticity, Special Publication No. 21. The Chemical Society, London 1967, p. 217.
2. R.B. Woodward and R. Hoffmann, The conservation of orbital symmetry. *Angew. Chem. Int. Ed.* **1969**, *8*, 781–853.
3. R.B. Woodward and R. Hoffmann, Stereochemistry of electrocyclic reactions. *J. Am. Chem. Soc.* **1965**, *87*, 395–397.
4. R. Criegee and K. Noll, Rearrangements in the 1,2,3,4-Tetramethylcyclobutane series. *Liebigs Ann. Chem.* **1969**, *627*, 1–14.
5. G.A. Doorakian and H.H. Freedman, An estimate of the relative rates of conrotatory vs disrotatory electrocyclic ring opening. *J. Am. Chem. Soc.* **1968**, *90*, 5310–5311.
6. E. Vogel, W. Grimme and E. Dinné, A study of the stereochemistry of the valence isomerization of trienes with central cis-double bonds to cyclohexa-1,3-dienes. *Tetrahedron Lett.* **1965**, *6*, 391–395.
7. D.S. Glass, J.W.H. Watthey and S. Winstein, Isolation and valency isomerization of cis-cis-cis-1,3,5-cyclononatriene. *Tetrahedron Lett.* **1965**, *6*, 377–383.
8. R. Huisgen, A. Dahmen and H. Huber, Specific conrotatory valence isomerization of octatetraenes to cycloocta-1,3,5-trienes. *J. Am. Chem. Soc.* **1967**, *89*, 7130–7131.
9. S.J. Cristol, R.M. Sequeira and C.H. DePuy, Bridged polycyclic compounds. XXXI. Stereochemical aspects of the solvolysis of cyclopropyl chlorides. *J. Am. Chem. Soc.* **1965**, *87*, 4007–4008.
10. P.S. Skell, S.R. Sandler, Reactions of 1,1-dihalocyclopropanes with electrophilic reagents. Synthetic route for inserting a carbon atom between the atoms of a double bond. *J. Am. Chem. Soc.* **1958**, *80*, 2024–2025.
11. R. Hoffmann and R.B. Woodward, Selection rules for concerted cycloaddition reactions. *J. Am. Chem. Soc.* **1965**, *87*, 2046–2048.
12. C-Y. Liu, D.A. Smith and K.N. Houk, An intramolecular [8+6] cycloaddition. *Tetrahedron Lett.* **1986**, *27*, 4881–4884.
13. R.C. Cookson, B.V. Drake, J. Hudec and A. Morrison, The adduct of tropone and cyclopentadiene: a new type of cyclic reaction. *Chem. Commun.* **1966**, 15–16.
14. H. Prinzbach and H. Knöfel, 7,9-Di-tert-butylsesquifulvalene: its synthesis and cycloadditions. *Angew. Chem. Int. Ed.* **1969**, *8*, 881–882.

15. W. von E. Doering, personal communication to Woodward cited in ref. 2.

16. H. Meister, The preparation of 1,8-dialkoxyocta-1,3,5,7-tetraenes and their conversion into 7,8-dialkoxybicyclo[4.2.0]octa-2,4-diene. *Chem. Ber.* **1963**, *96*, 1688–1696.

17. R.B. Woodward and R. Hoffmann, Selection rules for sigmatropic reactions. *J. Am. Chem. Soc.* **1965**, *87*, 2511–2513.

18. A.C. Cope and E.M. Hardy, The introduction of substituted vinyl groups. V. A rearrangement involving the migration of an allyl group in a three-carbon system. *J. Am. Chem. Soc.* **1940**, *62*, 441–444.

19. W.R. Roth, J. König and W. Stein, The stereochemistry of sigmatropic 1,5-hydrogen shifts. *Chem. Ber.* **1970**, *103*, 426–439.

20. H. Jiao and P.v.R. Schleyer, A detailed theoretical analysis of the 1,7-sigmatropic hydrogen shift: the Möbius character of the eight-electron transition structure. *Angew. Chem. Int. Ed.* **1993**, *12*, 1763–1765.

21. M. Akhtar and C.J. Gibbons, A convenient synthesis of vitamin D_3-9,19-^{3}H and the mechanism of the previtamin D3 vitamin D3 reaction. *Tetrahedron Lett.* **1965**, *6*, 509–512.

22. J.E. Baldwin and V.P. Reddy, Kinetics and deuterium effects for the thermal [1,7] sigmatropic rearrangements of *cis,cis*-1,3,5-octatriene. *J. Am. Chem. Soc.* **1987**, *109*, 8051–8056.

23. V.A. Mironov, E.V. Sobolev and A.N. Elizarova, Some general characteristic properties of substituted cyclopentadienes. *Tetrahedron* **1963**, *39*, 1939–1958.

24. W.R. Roth, 1,5-Hydrogen shifts in cyclopentadienes and indenes. *Tetrahedron Lett.* **1964**, *5*, 1009–1013.

25. G. Wilkinson and T.S. Piper, Alkyl and aryl derivatives of π-cyclopentadienyl compounds of chromium, molybdenum, tungsten and iron. *J. Inorg. Nucl. Chem.* **1956**, *3*, 104–124.

26. M.J. Bennett, Jr., F.A. Cotton, A. Davison, J.W. Faller, S.J. Lippard and S.M. Morehouse, Stereochemically nonrigid organometallic compounds. I. π-cyclopentadienyliron dicarbonyl σ-cyclopentadiene. *J. Am. Chem. Soc.* **1966**, *88*, 4371–4376.

27. P. Jutzi, Fluxional η^1-cyclopentadienyl compounds of main-group elements. *Chem. Rev.* **1986**, *86*, 983–996.

28. F.A. Cotton, A. Musco and G. Yagupsky, Stereochemically nonrigid organometallic compounds. VIII. Further studies of σ-cyclopentadienylmetal compounds. *J. Am. Chem. Soc.* **1967**, *89*, 6136–6139.

29. M. Stradiotto, D.W. Hughes, A.D. Bain, M.A. Brook and M.J. McGlinchey, The fluxional character of $(\eta^5$-$C_5H_5)Fe(CO)_2(\eta^1$-$C_9H_7)$: Evidence for the [4+2] cycloaddition of a metal-substituted isoindene with tetracyanoethylene. *Organometallics* **1997**, *16*, 5563–5568.

30. M. Stradiotto, S.S. Rigby, D.W. Hughes, M.A. Brook, A.D. Bain and M.J. McGlinchey, A multidimensional NMR study of tris(indenyl)methylsilane:

Molecular dynamics mapped onto a hypercube. *Organometallics* **1996**, *15*, 5645–5652.

31. A. Bonny, R.D. Holmes-Smith, G. Hunter and S.R. Stobart, Stereochemically nonrigid silanes, germanes and stannanes. 9. Chiral silylcyclopentadienes and related compounds: Mechanistic and stereochemical definition of fluxional behavior. *J. Am. Chem. Soc.* **1982**, *104*, 1855–1859.

32. S.S. Rigby, H.K. Gupta, N.H. Werstiuk, A.D. Bain and M.J. McGlinchey, The barriers to trimethylsilyl migrations in indenes and benzindenes: Silatropic shifts via aromatic transition states. *Polyhedron* **1995**, *14*, 2787–2796.

33. S.S. Rigby, H.K. Gupta, N.H. Werstiuk, A.D. Bain and M.J. McGlinchey, Do aromatic transition states lower barriers to silatropic shifts? A synthetic, NMR spectroscopic, and computational study. *Inorg. Chim. Acta* **1996**, *251*, 355–364.

34. S.S. Rigby, M. Stradiotto, S. Brydges, D.L. Pole, S. Top, A.D. Bain and M.J. McGlinchey, Diels-Alder dimerization of cyclopenta[l]phenanthrene (dibenz[e,g]indene) with iso-dibenzindene: A computational, NMR spectroscopic, X-ray crystallographic study. *J. Org. Chem.* **1998**, *63*, 3735–3740.

35. S. Rigby, L. Girard, A.D. Bain and M.J. McGlinchey, The molecular dynamics of d,l- and meso-bis(indenyl)dimethylsilane: A re-examination of the mechanism of interconversion by using single selective inversion NMR. *Organometallics* **1995**, *14*, 3798–3801.

36. A.D. McMaster and S.R. Stobart, Stereochemically nonrigid silanes, germanes and stannanes. 10. Diastereoisomerism and metallotropic behavior in polyindenyl derivatives of germanium and tin. Facile stereomutation. *J. Am. Chem. Soc.* **1982**, *104*, 2109–2112.

37. J.L. Atwood, A.D. McMaster, R.D. Rogers and S.R. Stobart, Stereochemically nonrigid silanes, germanes and stannanes. 12. Crystal and molecular structures of tetrakis(η^1-indenyl) derivatives of germanium and tin. Meso diastereomers with S_4 symmetry. *Organometallics* **1984**, *3*, 1500–1504.

38. M. Stradiotto, M.A. Brook and M.J. McGlinchey, The molecular dynamics and cycloaddition chemistry of tris(1-indenyl)allylsilane: Generation of the first crystallographically-characterised tris(benzonorbornyl)silane. *New J. Chem.* **1999**, 317–321.

39. M. Stradiotto, M.A. Brook and M.J. McGlinchey, The molecular dynamics and reactivity of tris(indenyl)silane: an NMR spectroscopic and X-ray crystallographic study. *J. Chem. Soc., Perkin Trans. 2* **2000**, 611–618.

40. M. Stradiotto and M.J. McGlinchey, η^1-Indenyl derivatives of transition metal and main group elements: synthesis, characterisation and molecular dynamics. *Coord. Chem. Rev.* **2001**, *219–221*, 311–378.

41. R.W. Murray and M.L. Kaplan, Sigmatropic reactions in the 1,4-bis(cycloheptatriene)benzene isomers. *J. Am. Chem. Soc.* **1966**, *88*, 3527–3533.

42. M.A. Battiste, Heptaphenyltropilidene — product of a Diels-Alder reaction of triphenylcyclopropene. *Chem. Ind.* **1961**, 550–551.

43. L.C.F. Chao, H.K. Gupta, D.W. Hughes, J.F. Britten, S. Rigby, A.D. Bain and M.J. McGlinchey, Chromium and molybdenum carbonyl complexes of C_7Ph_7H, $C_7Ph_5Me_2H$, and of $C_7Ph_7H(CO)$, the Diels-Alder adduct of tetracyclone and triphenylcyclopropene: A variable-temperature NMR and X-ray crystallographic study. *Organometallics* **1995**, *14*, 1139–1151.

44. H.K. Gupta, S. Brydges and M.J. McGlinchey, Diels-Alder reactions of 3-ferrocenyl-2,4,5-triphenylcyclopentadienone: Syntheses and structures of the sterically crowded systems C_6Ph_5Fc, C_7Ph_6FcH and $[C_7Ph_6FcH][SbCl_6]$. *Organometallics* **1999**, *18*, 115–122.

45. E.L. Mackor and C. MacLean, Intramolecular proton exchange in aromatic carbonium ions: isotope effect in hexameythylbenzene. *Pure Appl. Chem.* **1964**, *8*, 394–404.

46. W. von E. Doering and W.R. Roth, Thermal rearrangements. *Angew. Chem. Int. Ed.* **1952**, *2*, 115–122.

47. E. Vogel, K.-H. Ott and K. Gajek, Small carbon rings, VII. Valence isomerisation of cis-1,2-divinylcycloalkanes. *Liebigs Ann.* **1961**, *644*, 172–188.

48. J.M. Brown, B.T. Golding and J.J. Stofko, Jr., Isolation and characterisation of cis-divinylcyclopropane. *J. Chem. Soc. Chem. Commun.* **1973**, 319–320.

49. W. von E. Doering and W.R. Roth, A rapidly reversible degenerate Cope rearrangement: bicyclo[5.1.0]octa-2,5-diene. *Tetrahedron* **1963**, *19*, 715–737.

50. H.E. Zimmerman and G.L. Grundevald, The chemistry of barrelene III. A unique photoisomerization to semibullvalene. *J. Am. Chem. Soc.* **1966**, *88*, 183–184.

51. G. Schröder, Synthesis and properties of tricyclo[3.3.2.0$^{4.6}$]decatriene-(2.7.9) (Bullvalene). *Chem. Ber.* **1964**, *97*, 3140–3149.

52. R. Merényi, J.F.M. Oth and G. Schröder, The temperature-dependent NMR spectra of tricyclo[3.3.2.0$^{4.6}$]decadiene-(2.7) one of its derivatives and of tricyclo[3.3.2.0$^{4.6}$]decatriene-(2.7.9) (Bulllvalene). *Chem. Ber.* **1964**, *97*, 3150–3161.

53. J.F.M. Oth, K. Müllen, J.-M. Gilles and G. Schröder, Comparison of ^{13}C and ^{1}H magnetic resonance resonance spectroscopy as techniques for the quantitative investigation of dynamic processes. The Cope rearrangement in bullvalene. *Helv. Chim. Acta* **1974**, *57*, 1415–1433.

54. T. Koritsanszky, J. Buschmann, and P. Luger, Topological analysis of experimental electron densities. 1. The different C-C bonds in bullvalene. *J. Phys. Chem.* **1996**, *100*, 10547–10553.

55. K. Hafner, H.J. Lindner, W. Luo, K.-P. Meinhardt and T. Zink, Novel pericyclic reactions in π-perimeter chemistry. *Pure Appl. Chem.* **1993**, *65*, 17–25.

56. B.R. Beno, J. Fennen, K.N. Houk, H.J. Lindner and K. Hafner, [5,5] Sigmatropic rearrangement. DFT prediction of a diradical mechanism for a Woodward-Hoffmann "allowed" thermal pericyclic reaction. *J. Am. Chem. Soc.* **1998**, *120*, 10490–10493.

57. E.M. Greer and R. Hoffmann, Metalla-Cope rearrangements: bridging organic and inorganic chemistry. *J. Phys. Chem. A* **2010**, *114*, 8618–8624.

58. Y.-H. Tian, J. Huang and M. Kertesz, Fluxional σ-bonds of 2,5,8-tri-*tert*-butyl-1,3-diazaphenalenyl dimers: stepwise [3,3], [5,5] and [7,7] sigmatropic rearrangements *via* π-dimer intermediates. *Phys. Chem. Chem. Phys.* **2010**, *12*, 5084–5093.

59. K. Uchida, Z. Mou, M. Kertesz and T. Kubo, Fluxional σ-bonds of the 2,5,8-trimethylphenalenyl dimer: direct observation of the sixfold σ-bond shift via a π-dimer. *J. Am. Chem. Soc.* **2016**, *138*, 4665–4672.

60. Y. Morita, T. Aoki, K. Fukui, S. Nakazawa, K. Tamaki, S. Suzuki, A. Fuyihuro, K. Yamamoto, K. Sato, D. Shiomi, A. Naito, T. Takui and K. Nakasuji, A new trend in phenalenyl chemistry: a persistent neutral radical 2,5,8-tri-*tert*-butyl-1,3-diazaphenalenyl, and the excited triplet state of the Gable *syn*-dimer in the crystal of column motif. *Angew. Chem. Int. Ed.* **2002**, *41*, 1793–1796.

Chapter 4

Symmetry Breaking in Reaction Mechanisms and Rearrangements: The Spectroscopic, X-ray Crystallographic and Computational Approach

"The right question is usually more important than the right answer."

— *Plato*

In contrast to the classic work discussed in Chapter 2, more recent studies on the elucidation of reaction mechanisms have taken advantage of the enormous advances in spectroscopic techniques and computer hardware and software. Thus, in most cases, these problems can be tackled on a minuscule scale with almost no wastage of valuable material. We here discuss a number of the more illustrative examples of symmetry breaking taken from the literature in recent decades in which the structures of the molecules, the mechanisms of their reactions, or the details of their rearrangements were revealed by use of infrared, NMR or EPR spectroscopy, mass spectrometry, or X-ray crystallography.

4.1. Alkene Metathesis

Alkene metathesis brings about a redistribution of fragments of alkenes by the cleavage and reformation of carbon-carbon double bonds, a process mediated by a number of molybdenum- and tungsten-based catalysts.[1,2] It was already of great importance in the petroleum industry and has now become a major academic endeavour. Early mechanistic proposals involved the intermediacy of symmetric species derived from pairwise combinations of alkenes. Among these suggestions (Scheme 4.1) were a metal-cyclobutane complex, **1**,[3-5] with undefined metal-ligand bonding characteristics, a metallacyclopentane, **2**,[6] and a metal-tetracarbene species, **3**.[7] These were all eventually discarded and superseded by the mechanism proposed by Chauvin (Figure 4.1) whereby the reaction of a metal carbene complex and an alkene forms a metallacyclobutane intermediate that can cyclo-eliminate to regenerate the original reactants, or to form a new metal-alkylidene and a different alkene.[8]

Among the elegant experiments involving symmetry breaking that were designed to test this proposal was the molybdenum- or tungsten-catalysed reaction of 2,2'-divinylbiphenyl, **4**, together with its terminally tetradeuterated counterpart, **5**. If the reaction were to proceed in a pairwise manner, as in Scheme 4.2, the mixture of ethylenes formed would be comprised only of C_2H_4 and C_2D_4, along with phenanthrene.[12]

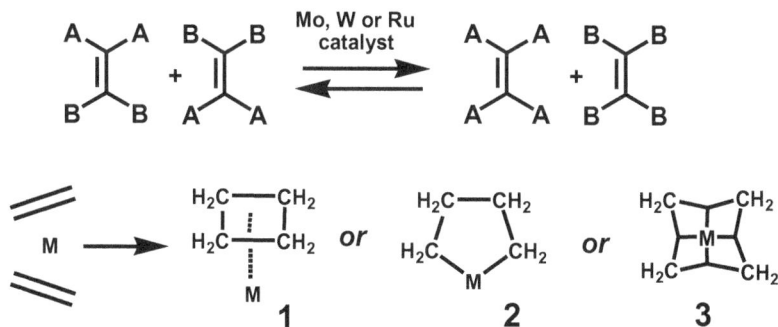

Scheme 4.1.   Alkene metathesis, and some early mechanistic proposals.

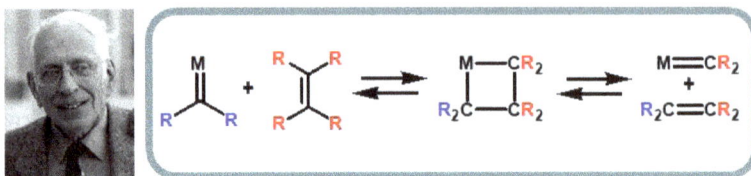

Figure 4.1. Yves Chauvin (1930–2015) of the Institut Français du Pétrole, and a member of the French Academy of Sciences, proposed the now-accepted metallacyclobutane mechanism for alkene metathesis. He shared the 2005 Nobel Prize in Chemistry with Richard Schrock (MIT) and Robert Grubbs (Caltech).[9-11]

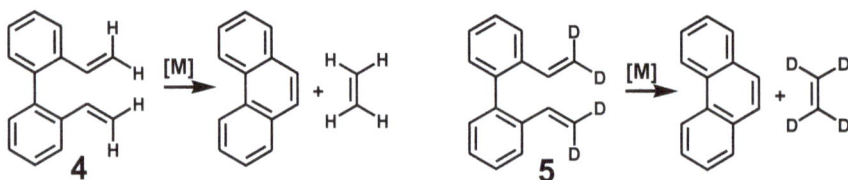

Scheme 4.2. The proposed pairwise metathesis mechanism, now known to be incorrect.

In contrast, the experimentally observed products, $H_2C=CH_2$, $H_2C=CD_2$ and $D_2C=CD_2$, in an approximate 1:2:1 ratio, are in complete accord with the Chauvin chain transfer mechanism via metallacyclobutane intermediates.[8] As delineated by Thomas Katz from Columbia University, New York, for the closely related 1,7-octadiene case, cycloaddition of the initiating $M=CR_2$ carbene complex forms the metallacyclobutane, 6, and elimination of $H_2C=CR_2$ generates the new carbene complex 7. Intramolecular metallacyclobutane formation, as in 8, followed by cycloreversion yields cyclohexene and liberates the methylene complex, $M=CH_2$, which can now add to terminally tetradeuterated 1,7-octadiene, as in 9, and eventually release $H_2C=CD_2$ (Scheme 4.3).[12]

Subsequently, a tungstenacyclobutane, 10, precisely of the kind proposed by Chauvin, was prepared in Schrock's group at MIT, and was unambiguously characterised by X-ray crystallography;[13] in

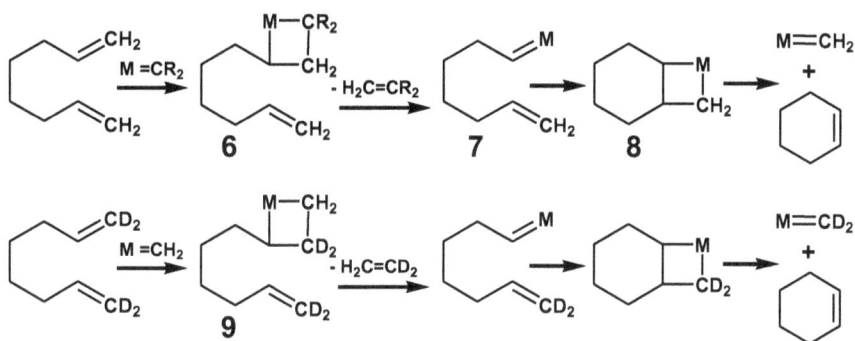

Scheme 4.3. Chauvin's metallacyclobutane mechanism satisfies all experimental observations.

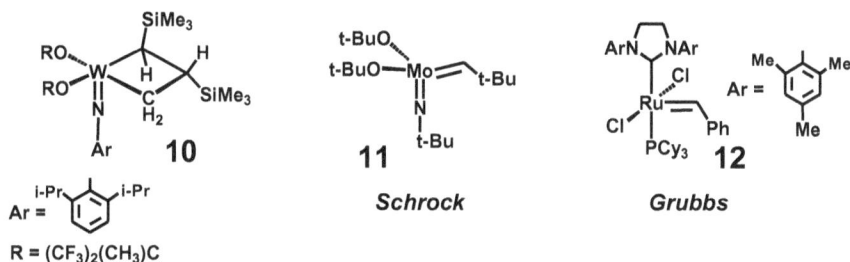

Figure 4.2. Schrock's metallacyclobutane, **10**, and Mo and Ru metathesis catalysts **11** and **12**.

this particular case, alkene elimination was relatively slow, but the analogous molybdenum systems with their weaker M–C bonds were markedly more reactive as metathesis catalysts. Typical catalysts currently in widespread use include the molybdenum-based system, **11**, reported by Schrock,[14] and the ruthenium complex, **12**, prepared by Grubbs at Caltech;[15] these are depicted in Figure 4.2.

This widely used reaction, especially for Ring-Opening Metathesis Polymerisation (ROMP) and Ring-Closing Metathesis (RCM) applications, is now a standard weapon in the synthetic chemist's armamentarium. Examples illustrating the preparation of a polymer and a natural product are shown in Scheme 4.4.

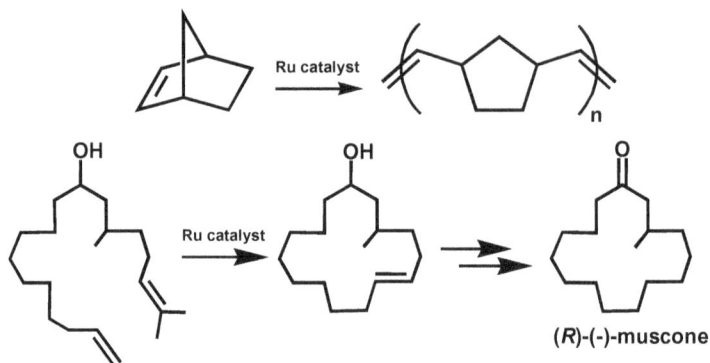

Scheme 4.4. ROMP of norbornene,[16] and RCM in the synthesis of (*R*)-(-)-muscone.[17]

4.2. Alkyne and Enyne Metathesis

In closely analogous work, alkyne metathesis proceeds via metallacy-clobutadiene intermediates and brings about a redistribution of fragments of alkynes by the cleavage and reformation of carbon-carbon triple bonds (Scheme 4.5).[18-20]

Once again, the contributions of the Schrock group have been pivotal. These include the isolation and X-ray crystallographic characterisation of a metallacyclobutadiene, **13** (Figure 4.3),[21] as confirmation of the mechanism, which has also been analysed in detail by calculations at the DFT level.[22]

Ring-closing alkyne metathesis has been used in spectacular fashion to prepare a wide range of natural products,[23] and also numerous macrocycles.[24] Typically, the cyclooligomers **14** and **15**, possessing three-fold and six-fold rotational symmetry, respectively, were prepared starting from *ortho* or *meta* aryl diynes,[25] with concomitant elimination of ethyne, as shown in Scheme 4.6.

Finally, brief mention should be made of enyne metathesis which provides a convenient route to 1,3-dienes, with stereochemical control.[26,27] The initial step involves cycloaddition of the initiating metal alkylidene to the alkyne thus forming a metallacyclobutene, **16**. Subsequent cycloreversion generates a double bond and a new metal

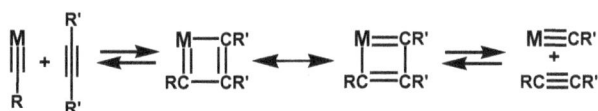

Scheme 4.5. Alkyne metathesis proceeds via a metallacyclobutadiene intermediate.

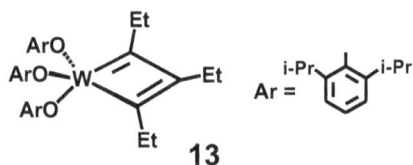

Figure 4.3. A tungstenacyclobutadiene complex characterised by X-ray crystallography.

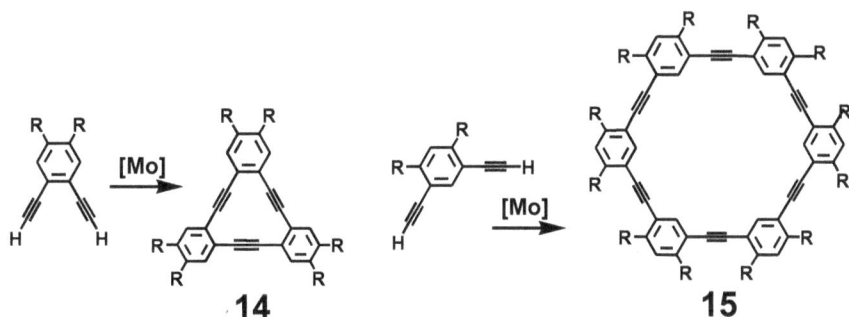

Scheme 4.6. Examples of cyclooligomers obtained via alkyne metathesis of aryl diynes.[25]

alkylidene, **17**, that undergoes conventional alkene metathesis ultimately furnishing a 1,3-diene (Scheme 4.7).

4.3. Carbonyl Insertions or Alkyl Migrations?

4.3.1. *^{13}CO-labelling studies on Manganese Carbonyl Complexes*

The reaction of methyl manganese pentacarbonyl, **18**, with triphenylphosphine yields complex **19** in which the phosphine is positioned

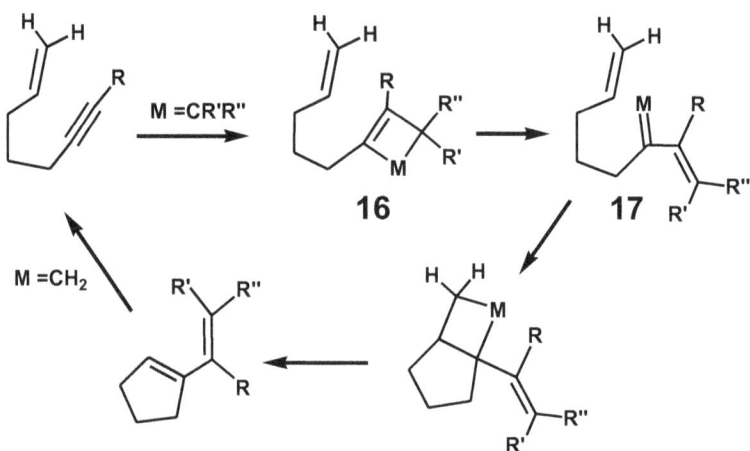

Scheme 4.7.   Enyne metathesis via a metallacyclobutane intermediate.

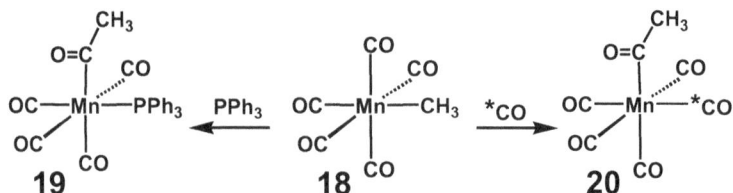

Scheme 4.8.   Formation of manganese-acyl complexes via methyl migration (*CO = ^{13}CO).

cis to the acyl substituent.[28] Likewise, when the incoming ligand is carbon monoxide, the product is $CH_3(C=O)Mn(CO)_5$, **20** (Scheme 4.8). This raised the mechanistic question as to whether this reaction occurs via direct insertion of the incoming carbon monoxide ligand into the metal-alkyl bond, or rather by alkyl migration with subsequent incorporation of CO.

The classic resolution of this problem was provided in an elegant infrared spectroscopic study by Fausto Calderazzo,[29] one of the great pioneers of mechanistic organometallic chemistry who led a group at the University of Pisa. One should note initially that the IR spectrum of locally C_{4v}-symmetric **20** in hexane exhibits three carbonyl stretching frequencies for the four in-plane ligands, at 2112 (A_1), 2048 (B) and 2006 cm^{-1} (E), another peak (A_1) at 2000 cm^{-1} for the single axial carbonyl, and finally at 1658 cm^{-1} for the acyl group.[30,31] When the complex is prepared from NaMn(CO)$_5$ and CH$_3$(^{13}CO)Cl, the acyl IR stretch is now found at 1625 cm^{-1}. When CH$_3$Mn(CO)$_5$ in heptane solution was carbonylated using ^{13}CO, the peaks attributable to the axial and acyl carbonyls were unchanged, but a peak at 1970 cm^{-1}, indicating the incorporation of ^{13}CO in an equatorial position, gradually increased in intensity. This showed unequivocally that the incoming ligand had not been incorporated into the acyl group.[30]

However, although this result clearly revealed the stereochemistry of the final product, it did not unambiguously distinguish between the scenarios whereby (a) the methyl migrated onto a *cis* carbonyl ligand prior to attack by the incoming ligand, or (b) that a *cis* carbonyl ligand moved so as to insert directly into the methyl-manganese bond. This dilemma was brilliantly resolved by decarbonylation (thermolysis at 55°C in hexane) of the acyl complex, **20**, in which one of the equatorial ligands was already labelled as ^{13}CO. The products were identified as the unlabelled material, **21**, and the *cis* and *trans* isomers, **22** and **23**, in the ratio 1:2:1, respectively (Scheme 4.9).[30] Evidently, the reaction proceeded by loss of an equatorial ligand with subsequent migration of the methyl group into the vacant site, thus demonstrating the microscopic reversibility of the carbonylation and decarbonylation processes.

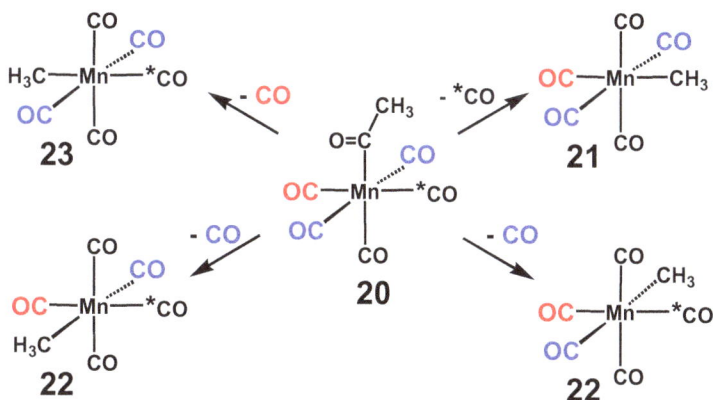

Scheme 4.9. Products from the decarbonylation of CH$_3$(C=O)Mn(CO)$_4$(^{13}CO).

This result paralleled an early observation by Coffield that upon decarbonylation the radiolabelled precursor $CH_3(^{14}C{=}O)Mn(CO)_5$ retained its ^{14}CO content.[32,33] Later work by Flood using ^{13}C NMR to monitor the reaction revealed that the methyl migration proceeded via a square-based pyramidal structure,[34] a conclusion also supported by a theoretical study.[35] Furthermore, it has been shown by using a chiral alkyl substituent that these migrations proceed with retention of stereochemistry.[36]

4.3.2. *^{13}CO-labelling Studies on Cobalt Carbonyl Clusters*

The chemistry of carbynyltricobalt nonacarbonyls, $RCCo_3(CO)_9$, has been investigated primarily at MIT by the group of Dietmar Seyferth, the distinguished founding editor of the journal *Organometallics*. These clusters are formed by the reaction of dicobalt octacarbonyl with compounds of the type $RCCl_3$, and have been characterised crystallographically as possessing a triangular C_{3v}-symmetric $Co_3(CO)_9$ unit capped by a carbynyl fragment that completes the tetrahedral framework.[37] Typical of this series are the clusters derived from CCl_4,[38] and from DDT,[39] **24** and **25**, respectively (Scheme 4.10).

Of particular interest was the report that, when the ester $Co_3(CO)_9CCO_2Me$ (readily obtainable from Cl_3CCO_2Me and $Co_2(CO)_8$) was treated with HPF_6 in propionic anhydride, the product was the acylium cluster, **26**,[41] analogous to those found upon treatment of 2,6-disubstituted aryl esters with strong acids **27→28** (Scheme 4.11).[42] Moreover, the acylium cluster was also formed upon treatment of the chloro derivative **24** with $AlCl_3$, whereby removal of the apical halogen brings about CO migration from a cobalt carbonyl

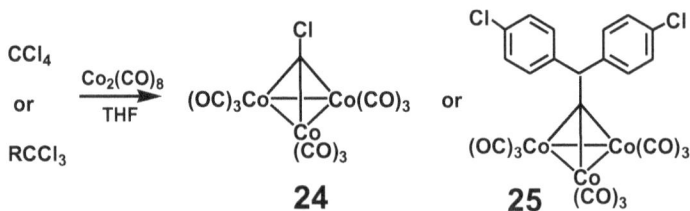

Scheme 4.10. Preparative route to carbynyltricobalt nonacarbonyl clusters.

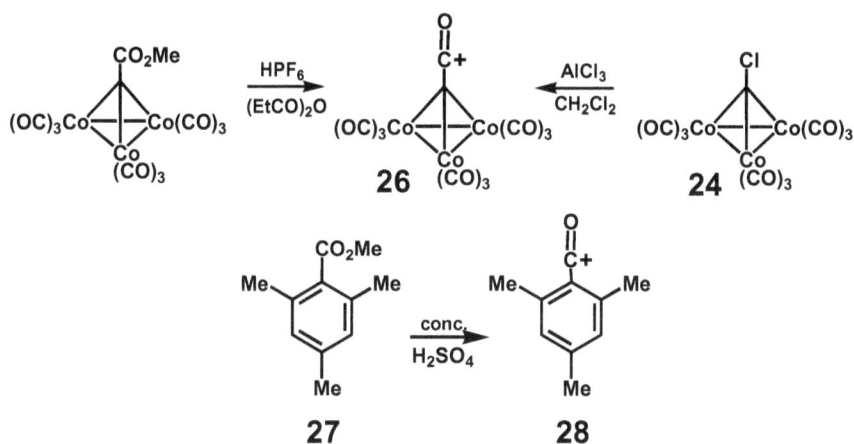

Scheme 4.11. Formation of organic and organometallic acylium cations.

position.[42] This latter route has the advantage that prior enrichment of the metal carbonyls with ^{13}CO leads to an enrichment of the ketenylidene carbon position with its attendant sensitivity enhancement for ^{13}C NMR observation.

The structure of the tricobaltcarbon decacarbonyl cation, **26**, has been probed by ^{13}C NMR spectroscopy which revealed a 6:3 splitting of the cobalt carbonyls at low temperature.[43] This observation might have been interpreted in terms of a C_{3v} structure, **26a**, exhibiting slowed exchange between the six equatorial and three axial ligands. However, since local rotation of each $Co(CO)_3$ vertex has always been seen to be fast on the NMR timescale, it was instead proposed that the capping acylium unit was tilted to adopt a mirror-symmetric (C_S) ketenylidene structure, **26b**, in which the positive charge is delocalised onto a cobalt vertex (Figure 4.4). Now, slowed migration of the ketenylidene fragment between the three cobalt vertices at low temperature can give rise to the observed 6:3 splitting, a scenario supported by molecular orbital calculations.[43] Moreover, Shriver has shown unequivocally by X-ray crystallography that in the isoelectronic anion $[Fe_3(CO)_9C=C=O]^{2-}$, **29**, the capping unit is tilted towards the triangular base.[44]

The carbonyl migration process **24→26** is matched by a somewhat unexpected decarbonylation. The acylium cation undergoes

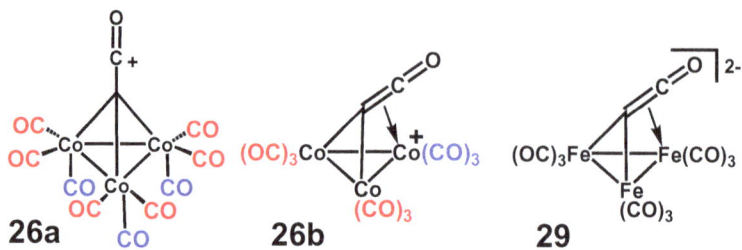

Figure 4.4.   Ketenylidene cations and anions of cobalt and iron carbonyls.

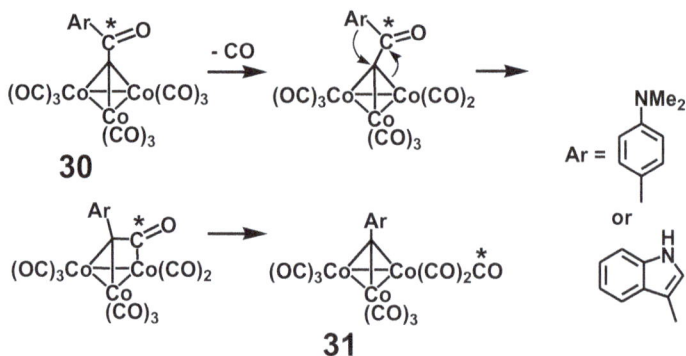

Scheme 4.12.   The decarbonylation mechanism in $Ar(C=O)Co_3(CO)_9$ clusters.

Friedel-Crafts reactions with electron-rich arenes to form ketones $Ar(C=O)CCo_3(CO)_9$, **30**; however, when gently heated these aroyl complexes suffer decarbonylation to the corresponding aryl deriva-tives, **31**.[45] Among the mechanistic possibilities, one might envisage a Norrish-type radical fragmentation process, but this is considered less probable because such reactions normally require photolysis, and the decarbonylation of **30** proceeds readily even in the dark. A more likely scenario involves decarbonylation via loss of a cobalt carbonyl ligand with subsequent migration of the ketonic moiety onto the vacant site on cobalt (Scheme 4.12). The latter mechanism would parallel the behaviour of the $CH_3(C=O)Mn(CO)_5$ system discussed above, and should be amenable to a similar ^{13}CO-labelling investigation. In the event, this is precisely what was observed when the cobalt carbonyls were enriched to the extent of 30% while the ketonic carbon remained at the 1% natural abundance level. Upon heating at reflux in benzene to bring about decarbonylation, the intensity of the cobalt carbonyl ^{13}CO resonance peak decreased relative to all others in the molecule by the expected factor of ~12.5%, or 1 in 8, clearly indicating the source of the ninth carbonyl in the triangular base.[46]

In a subsequent report, it was found that the formyl-$Co_3(CO)_7$(di-phosphine) complex, **32**, also suffered decarbonylation slowly even when stored at room temperature (Scheme 4.13), and the precursor formyl cluster and the carbynyl cluster, **33**, were both characterised by X-ray crystallography.[47] It is noteworthy that decarbonylation was not evident when the formyl complex was kept under an atmosphere of carbon monoxide.

One can now rationalise not only the decarbonylation of the aroyl-tricobalt nonacarbonyl cluster, **30**, but also the formation of the cation $[Co_3(CO)_9(C=C=O)]^+$, **26**, from $ClCCo_3(CO)_9$ and $AlCl_3$, in terms of a bridging carbonyl structure, as depicted in Scheme 4.14.

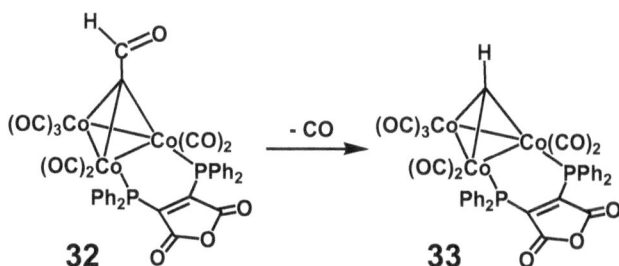

Scheme 4.13.   Decarbonylation of a formyl-tricobalt cluster.

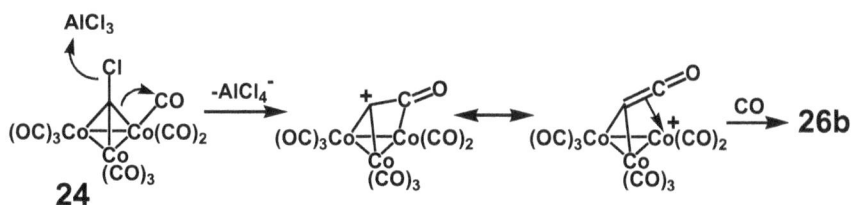

Scheme 4.14.   Migration of a CO ligand between a cobalt vertex and the capping carbon.

In each of these carbonylation/decarbonylation processes, the identity of the migrating group has been elucidated by breaking the symmetry of the system through the incorporation of ^{13}CO labels, and then monitoring their progress by using infrared or NMR spectroscopy.

4.3.3. *Alkyl Migration Leading to Metal Carbene Formation*

As part of his pioneering investigations of the chemistry of metal carbonyl anions, R.B. (Bruce) King, then at the Mellon Institute in Pittsburgh and subsequently a Distinguished Regents Professor at the University of Georgia, treated a series of dibromoalkanes and the corresponding acyl halides with $Na[(C_5H_5)Fe(CO)_2]$ and obtained the expected products of nucleophilic substitution, **34** and **35**, as shown in Scheme 4.15.[48,49] It was further noted that the acyl complex, **35**, was thermally stable even at 130°C, and contrasts with the well-known proclivity of acyl-$Mn(CO)_5$ complexes to lose a carbonyl ligand and undergo alkyl migration. King attributed this result to the enhanced Fe–CO bond strength relative to the Mn–CO case, a rationalisation supported by the stability of corresponding alkyl-rhenium systems. In subsequent publications from other groups, it was shown that these compounds could be interconverted under photolytic conditions,[50] and also that this reaction could be extended to include longer-chain dihaloalkanes ($n = 2–12$).[51]

In contrast, the reaction of $NaMn(CO)_5$ with 1,3-dibromopropane in THF yielded a material with the predicted formula,

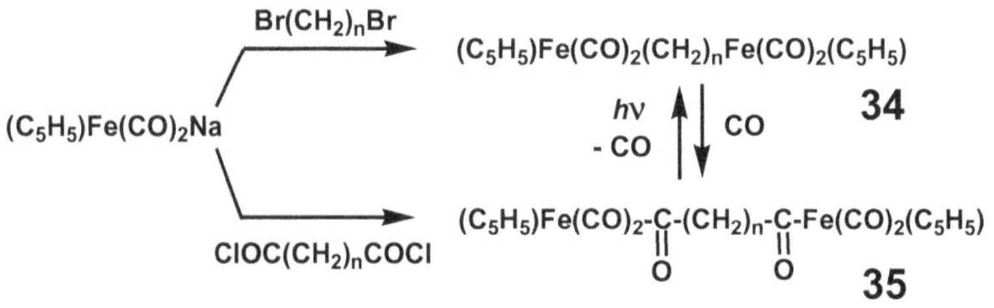

Scheme 4.15.   Reactions of $(C_5H_5)Fe(CO)_2Na$ with alkyl and acyl halides.

$(CH_2)_3[Mn(CO)_5]_2$, but with an unexpected ^1H NMR spectrum. The C_{2v} symmetry of the anticipated product $(OC)_5Mn\text{-}CH_2CH_2CH_2\text{-}Mn(CO)_5$, **36**, would normally lead one to expect a 1:2:1 triplet (4H) and 1:4:6:4:1 quintet (2H) peak intensity pattern in which the chemical shift of the central methylene is different from its outer counterparts. However, the ^1H NMR spectrum of the observed product appeared as a quintet and two non-equivalent triplets each of intensity corresponding to two hydrogens. In an attempt to rationalise this observation, it was suggested that the molecule adopted an unprecedented structure, **37**, whereby in each case the hydrogens in each of the pairs of methylene protons adjacent to manganese were non-equivalent, one of which interacted in an unspecified way with the more distant metal atom (Scheme 4.16).[52]

This project was revisited by C.P. Casey from the University of Wisconsin who reinterpreted the data in terms of an isomeric carbene structure, **38**, brought about by migration of the bromopropyl chain onto a *cis* carbonyl ligand whereupon nucleophilic attack at manganese by the second $Mn(CO)_5$ anion prompted ring closure with displacement of bromide (Scheme 4.17).[53,54] This structure accounted for the appearance of three non-equivalent methylene groups, and was later confirmed by X-ray crystallography.[55]

The molecule $(OC)_5Mn(CH_2)_3Mn(CO)_5$, **36**, was eventually successfully prepared by reaction of $KMn(CO)_5$ with the bistriflate $TfO(CH_2)_3OTf$.[56] We note also that analogous behaviour was found when $(C_5H_5)Mo(CO)_3Na$ reacted with the diiodoalkanes $I(CH_2)_nI$,

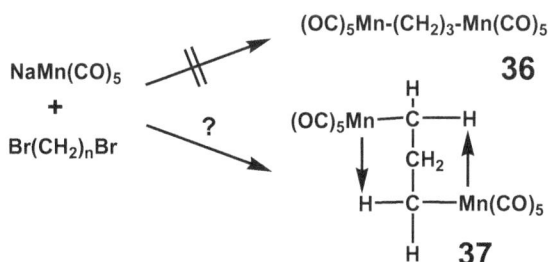

Scheme 4.16. King's initial proposal for the reaction of $NaMn(CO)_5$ with $Br(CH_2)_3Br$.

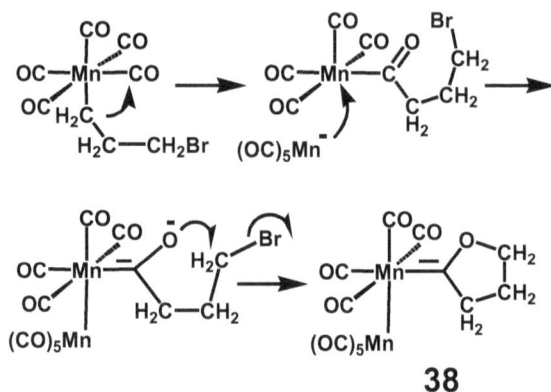

Scheme 4.17. Formation of carbene complex **38** by reaction of NaMn(CO)$_5$ with Br(CH$_2$)$_3$Br.

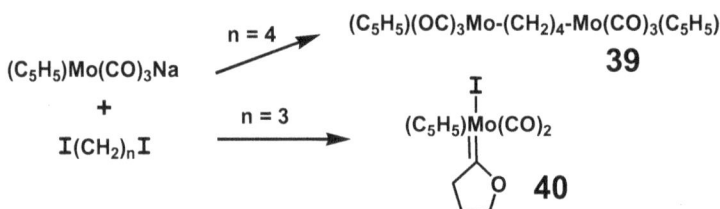

Scheme 4.18. Reactions of (C$_5$H$_5$)Mo(CO)$_3$Na with diiodoalkanes.

where n = 3 or 4. In the latter case, the product was the centrosymmetric compound (C$_5$H$_5$)Mo(CO)$_3$(CH$_2$)$_4$Mo(CO)$_2$(C$_5$H$_5$), **39**, whereas the reaction with diiodopropane yielded the carbene complex **40** (Scheme 4.18).[57]

4.4. Classical or Non-Classical Ions — The 2-norbornyl Cation Problem

In a "classical" carbocation the positive charge is localised on a single carbon atom or delocalised via resonance onto a neighbouring heteroatom or multiple bond such as in the allyl cation. In a "non-classical" carbocation the positive charge is delocalised by interaction with a non-adjacent single or multiple bond such that three or more atoms share a single pair of electrons. Bonding of this type is, of course, well characterised in boranes often designated as electron-deficient (even though the electron count is ideal for three-dimensional aromatic systems) and typified by icosahedral $[B_{12}H_{12}]^{2-}$ or square-based pyramidal B_5H_9.[58]

In early work by Winstein and Woodward, it was found that the rate of acetolysis of 7-norbornyl and *anti*-7-norbornenyl tosylates, **41** and **42**, respectively, differed by the enormous factor of 10^{11}; moreover, in the latter case the reaction proceeded with retention of configuration to yield only **43** (Scheme 4.19).[59] It was therefore suggested that the leaving group received anchimeric assistance from the double bond that partially stabilised the developing positive charge.

This proposal was strengthened by the observation that the ^1H NMR spectrum of the 7-norbornadienyl cation, **44**, generated by dissolution of 7-methoxynorbornadiene in fluorosulfuric acid, revealed the non-equivalence of the 2,3 and 5,6 positions, once again implying that the bridging carbocation was interacting preferentially with one face of the molecule. This was beautifully illustrated in an ingenious symmetry-breaking experiment to evaluate the barrier to ring flipping between isomers (Scheme 4.20). In the deuterium-labelled system,

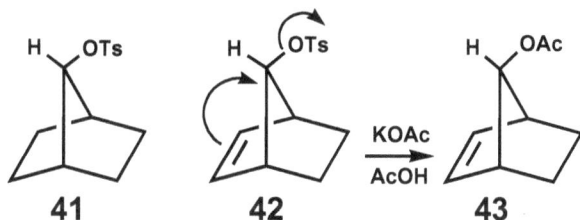

Scheme 4.19.   Acetolysis of 7-norbornyl tosylate and *anti*-7-norbornenyl tosylate.

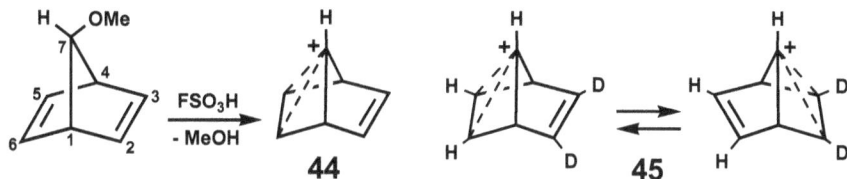

Scheme 4.20.   Flipping of the 7-norbornadienyl cation between the $C_2=C_3$ and $C_5=C_6$ bonds.

45, variable-temperature NMR measurements revealed a barrier of at least 19.6 kcal mol^{-1}.[60]

Subsequently, it was found that in the acetolysis of 2-norbornyl tosylate, **46**, an *exo/endo* rate difference of 350 was observed, and the product of nucleophilic attack was exclusively the *exo* isomer **47**.[61] Moreover, when this reaction was carried out on an enantiopure *exo* precursor, the product was fully racemised (Scheme 4.21).[62]

These data led Winstein to suggest that this again involved a non-classical carbocation whereby the positive charge was delocalised by interaction with a σ-bond, as in **48**, and the mirror symmetry of the non-classical cation led to racemisation. However, this proposal was vigorously disputed by Herbert C. Brown from Purdue University, Indiana, who argued that it was more readily rationalised in terms of a rapid equilibration between two classical carbocations **49** and **50** (Scheme 4.22).[63]

While many other researchers contributed usefully to the discussion,[64] it became a *cause célèbre* and unfortunately even led to *ad hominem* attacks. One should emphasise that Brown had no problem with the concept of delocalised electron-deficient bonding. He was a great pioneer in the field of boron chemistry, and indeed received the 1979 Nobel Prize for his work on hydroboration. In essence, the

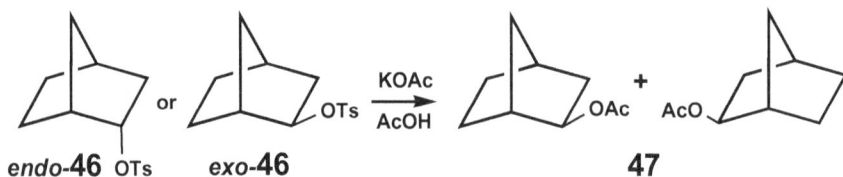

Scheme 4.21.   Racemisation of 2-norbornyl tosylate upon acetolysis.

Scheme 4.22.   Winstein's proposed mirror-symmetric non-classical species, **48**, and Brown's suggested rapid exchange between two classical carbocations, **49** and **50**.

issue came down to the establishment of whether the 2-norbornyl carbocation adopted a mirror-symmetric ground-state structure, or instead underwent rapid exchange between classical carbocations.

Around this time, George Olah entered the fray. He left Hungary after the revolution in 1956 and took up an industrial position at Dow Chemical in Canada, then became Chair of the Chemistry Department at Case Western Reserve University in Cleveland, Ohio, before eventually becoming Director of the Loker Hydrocarbon Research Institute at the University of Southern California. During this time, he pioneered the use of superacids, in particular the FSO_3H/SbF_5 combination (later known as Magic Acid) to prepare a wide range of long-lived stable carbocations and record their NMR spectra, for which he received the 1994 Nobel Prize in Chemistry.[65]

Early NMR studies on the 2-norbornyl cation were inconclusive because the system underwent numerous hydride shifts and Wagner-Meerwein rearrangements, and it was only with the advent of higher-field spectrometers (400 MHz) and the capacity to acquire spectra in unusual solvents, such as SO_2ClF, at very low temperature (115 K) that more definitive information became available. The simplicity of the ^1H and ^{13}C spectra indicating the mirror symmetry of the cation supported the non-classical model.[66,67] Even more impressively, Costantino Yannoni at the IBM Research Laboratory in San Jose, California was able to obtain the solid state ^{13}C NMR spectrum of the 1,2-dimethyl-2-norbornyl cation at 5 K;[68] once again there was no evidence for the existence of two individual classical carbocations and the result was in accord with a non-classical carbocation.

Nevertheless, since NMR is a relatively slow spectroscopic technique, the problem was also investigated by taking advantage of the much shorter timescale (picoseconds) of vibrational spectroscopy and the resulting Raman data again supported the existence of a single C_S-symmetric species rather than two separate classical carbocations.[69] In addition, use was made of X-ray photoelectron spectroscopy (also called Electron Spectroscopy for Chemical Analysis, ESCA), a technique that had been only recently developed at that time and involved the measurement of binding energies of core electrons, such as in the carbon $1s$ shell. The timescale of the measured ionisation processes is on the order of 10^{-16} s, much too fast to be averaged by any chemical exchange. The difference in $1s$ binding energy between a carbon clearly bearing a positive charge, as in the *tert*-butyl cation, and the other carbons is generally found to be in the range 4–5 eV. In the 2-norbornyl cation this difference is now only 1.7 eV, showing that the charge is not localised on a single carbon (or even on two rapidly equilibrating sites) but rather is dispersed over several positions in a non-classical structure.[70] Nevertheless, Brown continued to criticise the interpretation of these data.[71]

The gold standard, of course, would be an X-ray crystallographic structural determination of the 2-norbornyl carbocation, and in 1987 an attempt was made by Thomas Laube in Munich to obtain the structure of the 1,2,5,5-tetramethylnorbornyl system, **51**. Despite his valiant efforts, the cation underwent a multi-step Wagner-Meerwein rearrangement to give the 1,2,4,7-*anti*-tetramethylnorbornyl cation, **52**, the structure of which he succeeded in obtaining (Scheme 4.23).[72] Despite the failure to produce a potentially mirror-symmetric cation, the structure did reveal unsymmetrical σ-participation, in line with the Winstein model of non-classical behaviour.

This dispute was finally unambiguously resolved only in 2013 by the publication of the X-ray crystal structure, measured at 40 K, of the parent 2-norbornyl cation, **48**, with the $[Al_2Br_7]^-$ counterion, in which the non-classical structure (Figure 4.5) exhibited bond

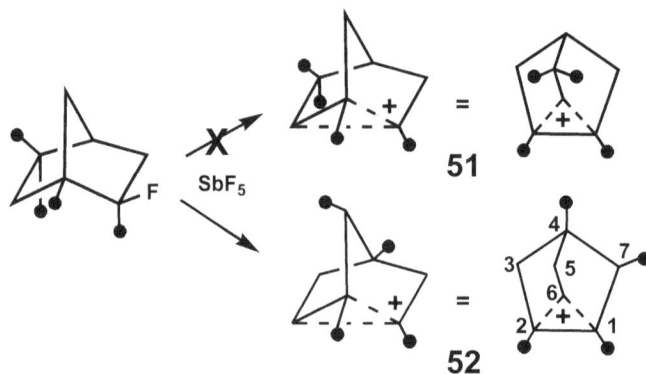

Scheme 4.23. Laube's X-ray crystal structure of the 1,2,4,7-*anti*-tetramethylnorbornyl.

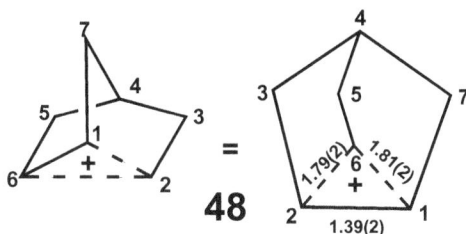

Figure 4.5. Selected structural parameters of the 2-norbornyl carbocation; distances in Å.

distances of ~1.80 Å from C(6) to C(1) and C(2) in three independent molecules within the unit cell.[73]

Perhaps the real significance of this work is that, given the problem of determining the symmetry of a particular molecular entity, the full gamut of spectroscopic, crystallographic and calculation techniques had to be invoked over a period of six decades to obtain an unambiguous result. Certainly, Brown forced researchers to consider the design of their experiments and their interpretation of the data with greater care and self-criticism.

4.5. The Curious Case of Racemisation During Nucleophilic Substitution in a (Fluoroarene)Cr(CO)$_3$ Complex

4.5.1. *The Preparation and Reactivity of Chiral Arene-chromium Carbonyl Complexes*

As noted in Chapter 1, breaking the mirror symmetry of the ring plane in an *ortho* or *meta* difunctionalised arene bearing different substituents generates enantiomers, as in **53**. We note that introduction of two different substituents into a five-membered ring in ferrocene, or of (cyclopentadienyl)Mn(CO)$_3$, also yields enantiomers.[74] Experimental verification of this concept in the (arene)Cr(CO)$_3$ case was provided by Gérard Jaouen, then at the University of Rennes, in the ancient capital of Britanny in north-western France, subsequently a distinguished professor in Paris. His group prepared chiral (π-arene) Cr(CO)$_3$ complexes (known in French as *benchrotrènes*) by this

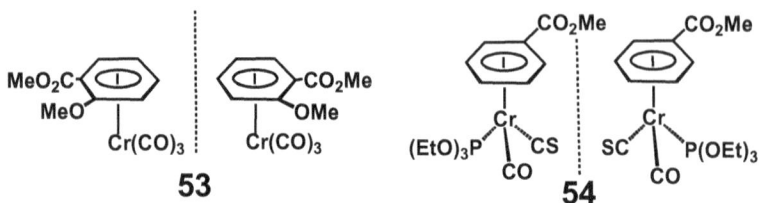

Figure 4.6. Examples of enantiomeric arene-chromium complexes: (i) by incorporation of two different substituents in the ring, **53**, or (ii) by making the tripod chiral, **54**.

particular approach,[75,76] but they also generated enantiomers by sequentially replacing two carbonyl groups by ligands such as thiocarbonyl or a phosphite, as in **54** (Figure 4.6).[77] Isolation of enantiopure products was accomplished by hydrolysis of the esters and treatment of the corresponding carboxylic acids with cinchonidine, or another chiral alkaloid, thus forming diastereomers that were separable by fractional crystallisation.[78]

Knowing that fluoro-substituted arenes readily undergo nucleophilic substitution when bearing strongly electron-withdrawing moieties such as nitro, cyano or, as in this case, a π-bonded $Cr(CO)_3$ unit,[79] in 1972 Jaouen prepared enantiomerically pure (methyl *o*-fluorobenzoate)chromium tricarbonyl, **55**, and treated it with sodium methoxide in refluxing methanol. To his surprise, the product (methyl *o*-methoxybenzoate) chromium tricarbonyl, **53**, was found to be approximately 50% racemised (Scheme 4.24).[80] Similar behaviour was found with other nucleophiles, such as diethylamine or sodium methanethiolate, but to a lesser extent. Interestingly, the corresponding *meta*-disubstituted complex apparently did not suffer racemisation.

If the only process operating were to be direct replacement of fluoride by nucleophilic attack at the *ipso* carbon via a Wheland-type intermediate, the molecule would retain its stereochemical integrity. However, it is clear that a competitive reaction leading to partial racemisation must intervene. One can envisage a number of mechanistic possibilities to rationalise these observations. The involvement of a benzyne intermediate, along with the possibility of complete detachment of the organometallic unit and reattachment to the opposite face of the arene, was readily dismissed. In the former case, one would

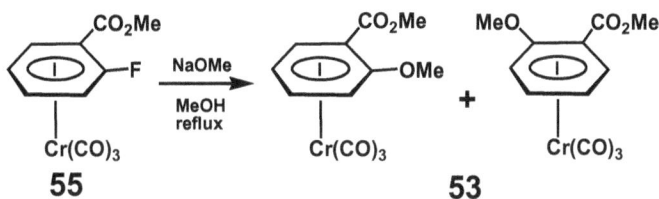

Scheme 4.24. Formation of partially racemised (methyl *o*-methoxybenzoate) $Cr(CO)_3$ during nucleophilic displacement of fluoride from enantiopure **55**.

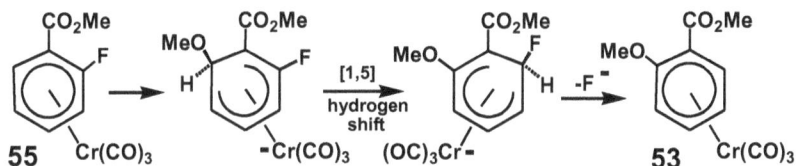

Scheme 4.25. A [1,5] hydrogen shift as a possible route to account for the racemisation of **53**.

expect to yield both *ortho* and *meta* products, and in the latter case, when the enantiopure starting material, **55**, was heated at reflux in methanol in the presence of free arene, its chiral character was unchanged. Clearly, the racemisation was intimately involved with the nucleophilic substitution process itself. Another possibility considered was that of partial detachment with the chromium linked only to the ester oxygen thus facilitating a "rollover" from one face of the arene to the other, but a more likely scenario involved nucleophilic attack at the "other" *ortho* position with subsequent [1,5] hydrogen migration, possibly via a chromium hydride, and then elimination of fluoride (Scheme 4.25).[80]

Some years later, important new observations relevant to this phenomenon were reported by Martin Semmelhack, then at Cornell, subsequently at Princeton. In the course of his studies on the nucleophilic replacement of the halide in (π-chlorobenzene)$Cr(CO)_3$ by the anion derived from isobutyronitrile, it was found by NMR monitoring during the course of the reaction that the anion attacked several positions on the arene ring. Furthermore, when the reaction mixture was treated with trifluoroacetic acid, and then with iodine to

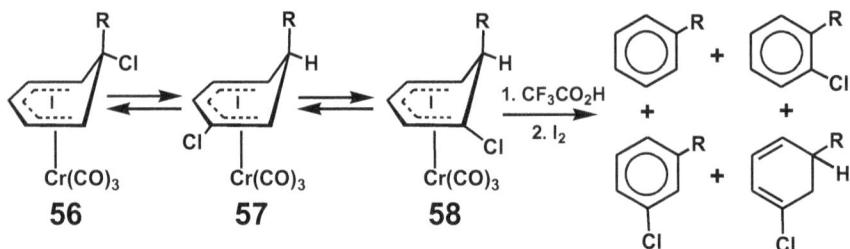

Scheme 4.26. A selection of arene or dihydroarene products from the reaction of (π-chlorobenzene)Cr(CO)$_3$ with the LiCMe$_2$CN after addition of CF$_3$CO$_2$H, and then iodine.

bring about decomplexation of the organometallic moiety, multiple products were identified as arenes containing either, or both, chlorine and the isobutyronitrile group, together with a number of dihydrobenzenes (Scheme 4.26).[81] These results were rationalised in terms of the generation of a mixture of Wheland-type intermediates, **56–58**, as are known to be formed in nucleophilic aromatic substitutions,[82] that can also suffer attack by acid to yield dihydroarenes.

4.5.2. *Cine, Tele-meta and Tele-para S$_{Ar}$N Substitutions in (arene)Cr(CO)$_3$ Systems*

Over several decades this area was extensively reinvestigated and, in 2002, was comprehensively reviewed by Françoise Rose-Munch and Eric Rose at Université Pierre et Marie Curie in Paris.[83] They noted that in very many cases nucleophilic attack did not occur only at the *ipso* carbon, but also at the positions *ortho*, *meta* or *para* to the leaving group. These are now designated as *cine*, *tele-meta*, and *tele-para* S$_{Ar}$N substitutions, respectively. (The *tele* nomenclature was introduced by IUPAC to denote reactions in which the entering group takes up a position further than one carbon away from the carbon to which the leaving group is attached.)

We here describe one representative experiment in which the mechanism has been clearly delineated via a combination of ^{1}H and ^{13}C NMR spectroscopy and deuterium labelling, and which provides a clear explanation for Jaouen's original observation. Treatment of

$(1,2,3$-trimethoxybenzene$)Cr(CO)_3$, **59**, initially with the lithium salts of CH_3CN, $MeCH_2CN$ or Me_2CHCN, and then with CF_3CO_2D, in each case led to displacement of a methoxide group with incorporation of the nucleophile to form either the *ipso* product, **60**, or the *tele-meta* isomer, **61**, as shown in Scheme 4.27. It is noticeable that increasing the steric bulk of the nucleophile markedly favours the *tele-meta* isomer.

As depicted in Scheme 4.28, initial attack at C(5) to form the Wheland intermediate, **62**, is followed by deuteration at chromium and migration onto the *endo* face of the ring in **63**. Next, a [1,5] migration of the hydrogen originally at C(5) *via chromium* onto C(3), **64→65**, generates the structure poised for elimination of MeOD and formation of the experimentally observed product **66**.[84]

In light of these data, one can now rationalise Jaouen's original report of racemisation[80] in terms of *cine* attack by methoxide at the

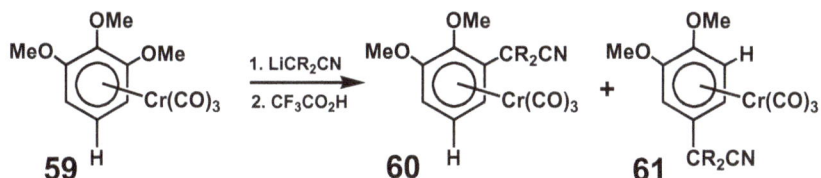

Scheme 4.27. Reactions of cyanoalkyl anions with $(1,2,3$-trimethoxybenzene$)$ $Cr(CO)_3$.

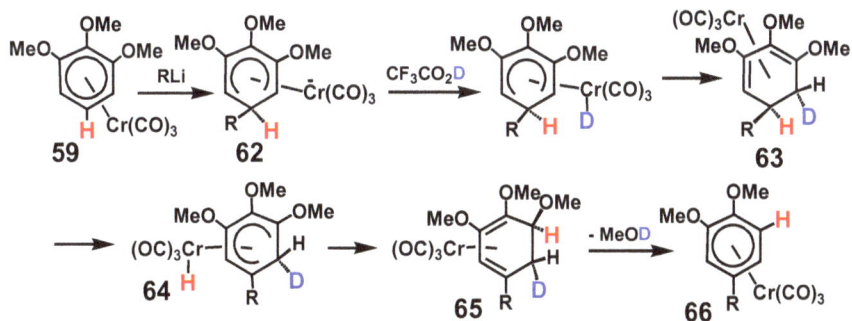

Scheme 4.28. Proposed mechanism for *tele-meta* nucleophilic attack on $(1,2,3$-tri-methoxy benzene$)Cr(CO)_3$, where $RLi = Me_2C(CN)Li$.

"other" *ortho* position with subsequent [1,5] hydrogen migration, via a chromium hydride, and elimination of fluoride, as was suggested in Scheme 4.25. Overall, the observed loss of planar chirality in **53** led ultimately to the elucidation of the mechanism of a previously unrecognised rearrangement.

4.6. Isotopically Chiral Tripods

4.6.1. *The Chiral Phosphoryl Group*

Many enzymatic reactions involve transfer of a phosphoryl group, but in 1975 the details of such processes had not been fully elucidated. In particular, the mechanism of phosphoryl transfer between adenosine triphosphate (ATP) and phosphoglycerate remained an open question: did this proceed via a direct transfer or rather involved an intermediate phosphorylated enzyme? The unambiguous establishment of the stereochemical course of such processes required the ready availability of a chiral probe to determine whether they occurred with inversion, retention or racemisation.

This was accomplished by the group led by Jeremy Knowles at Harvard who successfully prepared chiral phosphate monoesters by virtue of the incorporation of three isotopes of oxygen, as in $R-P(^{16}O,^{17}O,^{18}O)$. In the first of such reports, the route illustrated in Scheme 4.29 was described. Treatment of $^{17}O=PCl_3$ with (-)-ephedrine gave epimeric chloro adducts, **67**, that reacted with 2-benzyl-(*S*)-propane-1,2-diol to give the corresponding phosphoramidate esters, of which the major isomer, **68**, was obtainable in pure form after chromatographic separation. Subsequent ring opening with $H_2^{18}O$ to give **69** was followed by catalytic hydrogenolysis that furnished the desired $[1(R)-^{16}O,^{17}O,^{18}O]$phospho-(*S*)-propane-1,2-diol, **70**.[85] This was shown to be enantiomerically pure by a multistage process involving the mass spectrometric analysis of metastable ions; subsequently, a more convenient approach using ^{31}P NMR to determine the absolute configuration of these chiral phosphate groups was developed.[86]

With the chiral product in hand, Knowles was able to show that transphosphorylation from phenyl[(*R*)-$^{16}O,^{17}O,^{18}O$]phosphate to (*S*)-propane-1,2-diol, catalysed by alkaline phosphatase, proceeded with retention at phosphorus (Scheme 4.30).[87] This result is

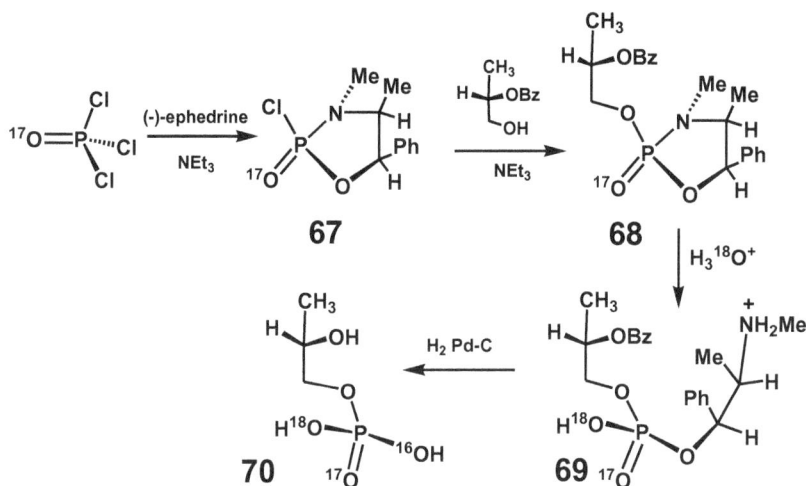

Scheme 4.29.   Synthesis of $[1(R)\text{-}^{16}\text{O},^{17}\text{O},^{18}\text{O}]$phospho-$(S)$-propane-1,2-diol, **70**.

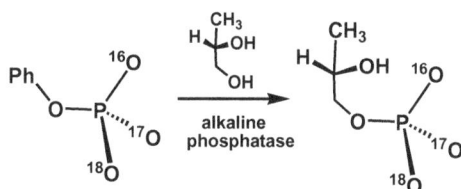

Scheme 4.30.   Enzyme-catalysed transfer of the chiral phosphoryl group to (S)-propane-1,2-diol with overall retention of configuration.

consistent with the formation of a phosphorylated enzyme intermediate, whereby retention is the result of two single displacement steps, each of which proceeds with inversion.

Since that time, other synthetic routes have been developed, and this approach has been widely used to study the stereochemical analyses of additional mutases, kinases and phosphatases, as exemplified in an important review by Gordon Lowe from Oxford.[88]

4.6.2. *The Chiral Methyl Group*

Prior to the elegant studies of Knowles and Lowe labelling phosphates with three isotopes of oxygen, a number of researchers chose

Figure 4.7. Sir John Cornforth (1917–2013), Nobel Prize 1975, and the Google doodle celebrating the 100th anniversary of his birth.

to study what must surely have been the ultimate in stereogenic discrimination — the chiral methyl group, $C(^1H,^2H,^3H)$. The original pioneers adopted different approaches: Arigoni prepared R and S enantiomers independently by successive stereospecific introduction of the three isotopes of hydrogen at the same carbon atom.[89] In contrast, the Cornforth group (Figure 4.7) synthesised racemic compounds of the type $R^*C(^1H,^2H,^3H)$ in which the configuration of the chiral substituent R^* was correlated with that of the chiral methyl group. Resolution of these enantiomers separated the R and S methyl species.[90,91]

Although numerous approaches have since been adopted, perhaps the most aesthetically satisfying is that reported by Duilio Arigoni at ETH Zürich whereby a spectacular cascade of electrocyclic reactions positions the three isotopes on the same carbon stereospecifically in a single process.[92] As shown in Scheme 4.31, thermolysis of the propargylic ether, **71**, at 260°C in a sealed tube yielded 2-methyl-3-cyclohex-1-enylcyclopentene, **72**, with elimination of methyl formate.

Realising that this must have occurred via an ene reaction to produce **73**, with subsequent reductive elimination of methyl formate, the authors decided to synthesise a chiral version of **71**, and also to incorporate deuterium at the methylene positions of the ether side chain. Finally, treatment with tritiated water and a trace of base brought about replacement of the alkynyl hydrogen to furnish the required precursor, **74**. Thermolysis as before yielded **75**

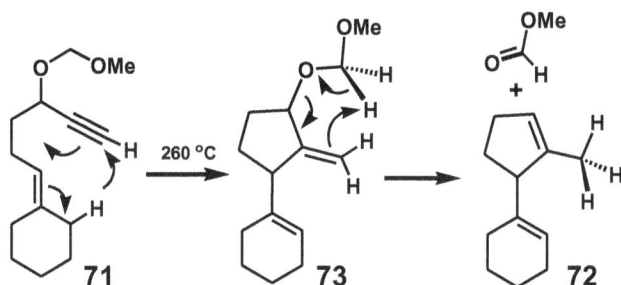

Scheme 4.31. Thermolytic rearrangement of a propargyl ether, **71**.

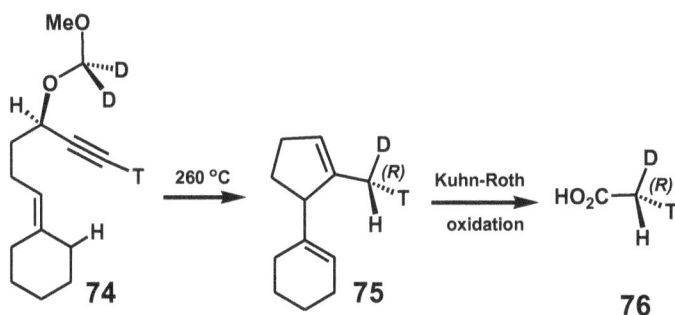

Scheme 4.32. Arigoni's route to a chiral methyl group.

bearing the chiral methyl group, and Kuhn-Roth oxidation delivered chiral acetic acid, **76** (Scheme 4.32).[92] The configuration of the generated chiral methyl group is imposed by the absolute configuration of the starting material, **74**, and by the strict geometric requirements imposed by the two transition states leading to the final product.

Subsequently, an efficient synthesis was developed by the group of Heinz Floss at Purdue University, Indiana, by using more conventional chemistry (Scheme 4.33). This involved the reduction of the deuterated dimethoxybenzaldehyde, **77**, to form the benzyl alcohol, **78**, in which the primary alcohol grouping was isotopically chiral. Conversion to the corresponding tosylate, **79**, sets up the molecule for reaction with tritiated superhydride, LiEt$_3$BT, thereby generating the chiral methyl group in **80**. Ozonolysis and

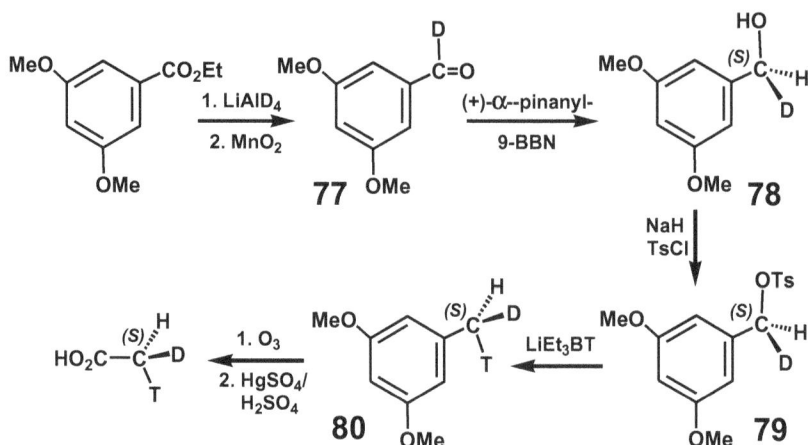

Scheme 4.33. Floss's route to a chiral methyl group.

Kuhn-Roth oxidation furnished isotopically chiral acetic acid in good yield, and 91% ee.[93]

As with the chiral phosphate system discussed above, one can use NMR for stereochemical studies involving chiral methyl groups, thus avoiding the need for chemical degradation. In this case, the pioneering work of Lawrence Altman's group at the State University of New York at Stony Brook on tritium NMR was crucial.[94] It is important to note that since ^3H nuclei resonate at 107.6 MHz (with ^1H at 100 MHz), J couplings to tritium are 7.6% larger than the corresponding couplings to protons. This is particularly noticeable in systems containing deuterium since, on a 100 MHz spectrometer, ^2H resonates at 15.36 MHz, so that $J(^3H–^1H)$ couplings are larger than the corresponding $J(^2H–^1H)$ values by a factor of 7!

The ready availability of such a molecular functionality has found numerous applications in biochemistry. Since it is now well established that single intermolecular methyl transfers proceed with inversion, one can now decide unequivocally whether any particular sequence of reactions proceeds via an odd or even number of methyl transfers.[95] Typically, as depicted in Scheme 4.34, the steric course of the enzymatic synthesis of methyl coenzyme M, **81**, from methanol in *M. barkeri* proceeds via two inversion steps. It was then degraded to acetate for configurational analysis.[96]

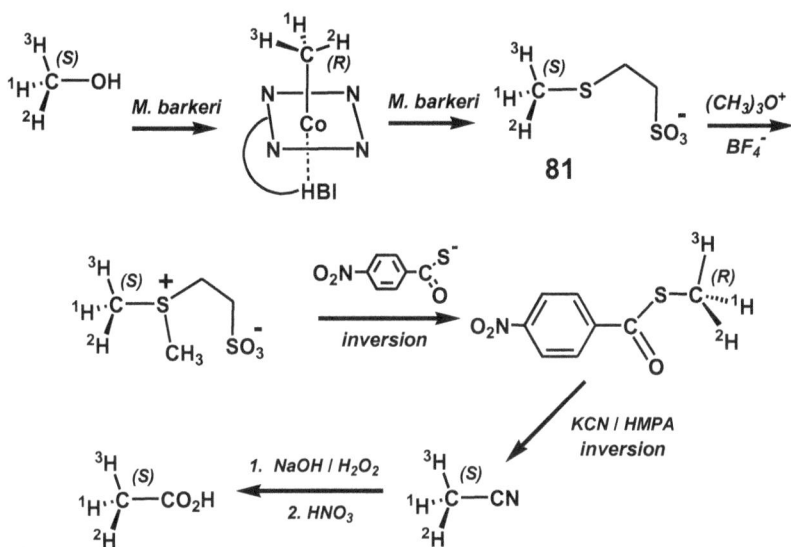

Scheme 4.34. Synthesis of methyl coenzyme M, **81**, from methanol in *M. barkeri* proceeds with overall retention of configuration (HBI = 5-hydroxybenzimidazole).

4.7. Use of Isotopes other than ^{2}H or $^{13/14}$C as Mechanistic Probes

4.7.1. *Electron Transfer Processes*

One of the apparently simple, yet actually rather sophisticated, processes in chemistry is that of electron transfer between transition metal complexes. The net result of the migration of a single electron from one metal to another increases the formal oxidation state of one while concomitantly lowering that of the recipient. It may be subdivided into two extreme scenarios: in one case an electron is transferred and the composition of each coordination sphere remains intact; this is designated as an *outer sphere* process. However, there is another class of reactions in which an atom or group is transferred from one metal to the other via a transient bridged species — an *inner sphere* electron transfer process; the changes in oxidation states after the transfer commonly bring about further reaction, generally with the solvent. These processes have been the topic of much profound analysis that led to Nobel awards for Henry Taube at Stanford in 1983, and for Rudolph (Rudy) Marcus at Caltech in 1992, both of whom were born in Canada (Saskatchewan and Montreal, respectively), and eventually moved to California.

4.7.2. *Outer Sphere Electron Transfer*

In electron transfer processes in which both participants are kinetically inert and ligand dissociation is very slow, reaction occurs by an outer sphere mechanism whereby electron density is delocalised onto the ligands of the donor metal and thence onto the recipient metal complex. As elucidated by Marcus,[97] and also by Noel Hush,[98] the distinguished Australian theoretician, in accord with the Franck-Condon principle the bond distances within the two systems have to adjust prior to the electron transfer to minimise the activation energy barrier. However, our emphasis here is not on these factors, important as they evidently are, but rather on the need for symmetry breaking to follow the progress of the reaction, and hence measure the kinetics of the electron transfer process.

In many such reactions, the identities of the reactants and products are the same (self-exchange reactions) and so a method of monitoring the extent of exchange over time is required; in a number of cases, this has been accomplished by use of NMR line-broadening techniques[99] as, for example, in the case of electron transfer between the Fe(II) and Fe(III) complexes of the tris(o-phenanthroline)iron cations, $[Fe(o\text{-phen})_3]^{2+}$ and $[Fe(o\text{-phen})_3]^{3+}$. However, we focus here on the ingenious approaches taken whereby one, or more, of the reactants is/are labelled either by isotopic substitution or by invoking chirality.

Typically, in the exchange between the kinetically inert low-spin d^6 Fe(II) and d^5 Fe(III) ferrocyanide and ferricyanide anions, $[Fe(CN)_6]^{4-}$ and $[Fe(CN)_6]^{3-}$, respectively, the symmetry needs to be broken without substantially altering the fundamental nature of the process. This was accomplished by the incorporation of a radioactive isotope of iron, ^{59}Fe, into one of the starting materials.[100]

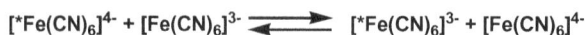

$$[^*Fe(CN)_6]^{4-} + [Fe(CN)_6]^{3-} \rightleftharpoons [^*Fe(CN)_6]^{3-} + [Fe(CN)_6]^{4-}$$

This approach was pioneered largely by Arthur Wahl, a remarkably talented scientist in whose PhD thesis at the University of California, Berkeley, under the guidance of Glenn Seaborg (Nobel Prize 1951), he reported for the first time the isolation and purification of plutonium. He continued his academic career in St Louis, Missouri, where access to the Washington University cyclotron allowed his group to generate a wide range of metal isotopes. As early as 1948, he described the kinetics of the exchange between thallous and thallic perchlorates by tagging the Tl(III) species with [204]Tl. In this experiment, after removal of aliquots for radiochemical assay, separation of the different oxidation states was accomplished by precipitation of Tl(I) as the chromate, while the Tl(III) fraction was precipitated as thallium hydroxide.[101] Analogously, the exchange between manganate, $[Mn^{VI}O_4]^{2-}$, and permanganate, $[Mn^{VII}O_4]^-$, ions was followed by labelling the Mn(VI) ion with radioactive [54]Mn. The permanganate was assayed after rapid coprecipitation with tetraphenylarsonium perrhenate, $[Ph_4As][ReO_4]$.[102,103] Likewise, electron transfer between vanadium(II) and vanadium(III) species was followed by incorporating [48]V as a tracer in the latter species.[104] Many other studies on a wide range of different elemental systems were carried out; the Ag(I)/Ag(II) system was investigated by using the radioisotope [110m]Ag,[105] and in a study of Sb(III)/Sb(V) exchange using [124]Sb as the tracer, separation of the products was achieved by extracting the antimony(V) species into diisopropyl ether from chloroform.[106] A different approach, electrochemical separation, was taken for the Ce(III)/Ce(IV) exchange using a [141]Ce label whereby the cerium(III) cation migrated to one electrode while the cerium(IV), as the hexanitritocerate anion $[Ce(NO_2)_6]^{2-}$, migrated to the other.[107]

A different and particularly ingenious approach to a related situation was taken by Francis Dwyer from the University of Sydney, Australia. In that case, rather than use a radioactive tag, he chose to take advantage of the very large optical rotations associated with some tris-chelate complexes of Group 8. As discussed in Chapter 1, in 1912 Alfred Werner had prepared the *l*-enantiomer of $[Fe(bipy)_3]^{2+}$ as the bromide and iodide and recorded their optical rotation, $[\alpha]^{20}_D$, as −520° and −440°, respectively.[108] Four decades later, Dwyer and

Gyarfas prepared and separated the corresponding perchlorate salts, via their Δ-tartrates, and found that their optical rotations, $[\alpha]^{20}{}_D$, were very much higher with values of ± 4800°.[109] These were then oxidised with ceric ammonium nitrate to furnish their Fe(III) counterparts. Similarly, they isolated the corresponding Λ- and Δ-osmium complexes, dark red $[Os(bipy)_3]^{2+}$ and dark green $[Os(bipy)_3]^{3+}$, respectively, as chloride salts, but found that they showed no detectable rotation at the sodium D line, the wavelength of which is too close to an absorption maximum. However, in mercury light (λ 5461 Å) the large $[\alpha]^{15}{}_{5461}$ value of 2100° was observed for the Os(II) salt.[110]

The availability of these $[Os(bipy)_3]^{2/3+}$ salts as single enantiomers prompted a study of their electron transfer behaviour, in the knowledge that intramolecular racemisation of either of these individual cations would be extremely slow. Now, taking equimolar quantities of Λ-$[Os(bipy)_3]^{2+}$ and Δ-$[Os(bipy)_3]^{3+}$ brought about electron transfer from the Os(II) to the Os(III) species thereby generating increasing amounts of Λ-$[Os(bipy)_3]^{3+}$, ultimately resulting in formation of racemic $[Os(bipy)_3]^{3+}$, and of course racemic $[Os(bipy)_3]^{2+}$ by the reverse process. Thus, the progress of the reaction depicted in Scheme 4.35 could be conveniently followed by polarimetry, and yielded a rather slow k value of $\geq 5 \times 10^{-4}$ M^{-1}s^{-1}.[111]

4.7.3. *Inner Sphere Electron Transfer*

Interactions between metal complexes such that one is kinetically inert and the other is labile can proceed via the atom transfer route.

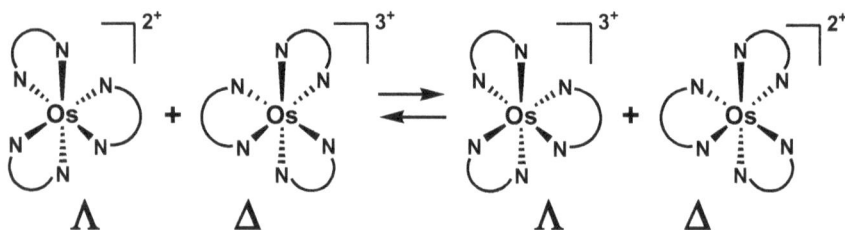

Scheme 4.35. Racemisation of $[Os(bipy)_3]^{2/3+}$ to monitor the rate of electron transfer.

A famous, now classic, experiment was described by Taube in which an electron is transferred from a labile high-spin d^4 hexaaquochromium(II) ion to an inert low-spin d^6 chloropentamminecobalt(III) complex, thus yielding a substitution-inert d^3 Cr(III) complex and a labile high-spin d^7 Co(II) ion that undergoes further ligand displacement by reaction with the solvent.

The reaction proceeds by loss of an aquo ligand from the Cr(II) complex thus leaving a vacant coordination site which can be accessed by the chloride bonded to cobalt(III), thereby forming a transient chloride-bridged species, **82**, in which electron transfer from Cr(II) to Co(III) is now facile. The newly generated Cr(III) species is inert; in contrast, the labile Co(II) complex loses all the ammine ligands under acidic conditions (Scheme 4.36).[112]

Most importantly, from the symmetry-breaking perspective, when the reaction was carried out with ^{36}Cl-labelled chloride in the solution, none of the radioactivity was found in the $[Cr(H_2O)_5Cl]^{2+}$ product, demonstrating that the bridging chlorine atom was bonded to both metals throughout the transfer giving no opportunity for incorporation of external ^{36}Cl$^-$.

Multitudinous examples of electronic interactions between metals via bridging ligands possessing aromatic or other unsaturated linkages have been reported, but perhaps the most spectacular are those involving long polyalkyne units connecting two organometallic centres,[113] as in the complex $(C_5H_5)W(CO)_3-[C\equiv C]_n-Fe(CO)_2(C_5H_5)$, where $n = 2$,[114] or in $(C_5Me_5)Re(NO)(PPh_3)-[C\equiv C]_n-Re(NO)(PPh_3)$ (C_5Me_5), where $n = 2-10$.[115]

$$\boxed{[(H_2O)_5Cr \text{--} Cl \text{--} Co(NH_3)_5]^{4+}} \quad \mathbf{82}$$

$$[Cr(H_2O)_6]^{2+} + [Co(NH_3)_5Cl]^{2+} \longrightarrow [Cr(H_2O)_5Cl]^{2+} + [Co(NH_3)_5(H_2O)]^{2+}$$
$$d^4 \text{ labile} \qquad d^6 \text{ inert} \qquad\qquad\qquad d^3 \text{ inert} \qquad\qquad d^7 \text{ labile}$$

$$[Co(NH_3)_5(H_2O)]^{2+} \xrightarrow{5H_3O^+} [Co(H_2O)_6]^{2+} + 5NH_4^+$$

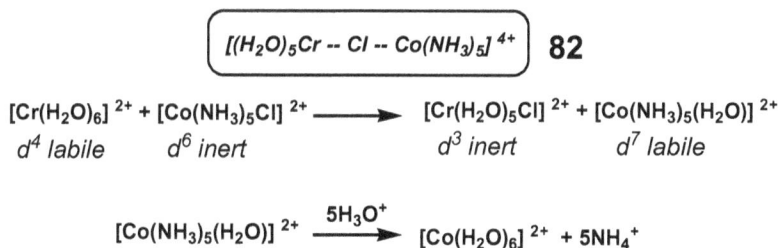

Scheme 4.36. Inner sphere electron transfer from Cr(II) to Co(III) via a chlorine bridge.

4.7.4. *A Classic Crossover Experiment Involving Metal-metal Bonds*

The mechanisms of substitution reactions of metal carbonyl dimers have been a long-time area of discussion. Since the average metal-metal and metal-carbonyl bond energies are comparable, the reactions could proceed either by loss of CO (Scheme 4.37), or via metal-metal bond cleavage (Scheme 4.38). To distinguish between these possibilities, Muetterties devised a crossover experiment using different isotopes of rhenium.[116] This element is found in nature as two isotopes, stable ^{185}Re (37.4%) and ^{187}Re (62.6%); the more abundant isotope is nominally a β-emitter, but has a half-life of 41.2×10^9 years, a time longer than the age of the universe. Therefore, the extent of crossover is most conveniently monitored by mass spectrometry.

Both isotopes of rhenium were available from Oak Ridge National Laboratory, Tennessee, in > 96% purity and were used to prepare ^{185}Re$_2$(CO)$_{10}$ and ^{187}Re$_2$(CO)$_{10}$. When a solution of these isotopomers, 50/50 in *n*-octane, was kept at room temperature in the dark for 16 h, the extent of crossover was less than 2%, as recorded by mass spectrometric analysis which revealed parent peaks at *m/e* 650 and 654. In contrast, when held in a vacuum-sealed tube at 150°C for 14 h under an argon atmosphere, or photolysed for 20 minutes using a medium-pressure mercury lamp, almost complete crossover was observed, as shown by the observed *m/e* values at 650, 652 and 654 in an

$$M_2(CO)_{10} \rightleftarrows M_2(CO)_9 + CO$$
$$M_2(CO)_9 + L \longrightarrow M_2(CO)_9L$$

Scheme 4.37. Ligand substitution involving reversible loss of carbon monoxide.

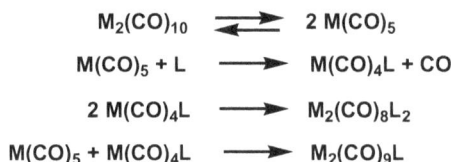

$$M_2(CO)_{10} \rightleftarrows 2\ M(CO)_5$$
$$M(CO)_5 + L \longrightarrow M(CO)_4L + CO$$
$$2\ M(CO)_4L \longrightarrow M_2(CO)_8L_2$$
$$M(CO)_5 + M(CO)_4L \longrightarrow M_2(CO)_9L$$

Scheme 4.38. Ligand substitution via metal-metal bond breaking.

$$^{185}Re_2(CO)_{10} + PPh_3 \longrightarrow {}^{185}Re_2(CO)_9(PPh_3) + CO$$

$$^{187}Re_2(CO)_{10} + PPh_3 \longrightarrow {}^{187}Re_2(CO)_9(PPh_3) + CO$$

$$^{185}Re^{187}Re(CO)_9(PPh_3) \text{ not formed}$$

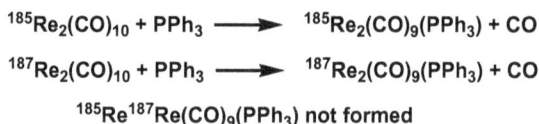

Scheme 4.39. Ligand substitution reactions using $^{185}Re_2(CO)_{10}$ and $^{187}Re_2(CO)_{10}$.

approximate 1:2:1 ratio. However, although thermolysis at 150°C for 14 h under an atmosphere of ^{13}CO only gave rise to background crossover (2%), incorporation of labelled carbon monoxide was very substantial. Likewise, reaction with triphenylphosphine also proceeded smoothly and, once again, crossover products were not formed (Scheme 4.39).

These results led Muetterties to conclude that the mechanism of ligand substitution under thermal conditions involved reversible CO dissociation, as in Scheme 4.38, and did not proceed via metal-metal bond breaking. In contrast, photochemical initiation of the reaction led to extensive Re–Re bond cleavage with consequent substantial crossover production.

4.8. Sulfur-nitrogen Ring Rearrangements

4.8.1. *Dynamic Behaviour of $[S_4N_5]^+$ and S_5N_6*

The chemistry of sulfur-nitrogen compounds with their novel reactivity, wide range of structures, and unique magnetic and conducting properties continues to attract attention.[117] As depicted in Figure 4.8, the D_{4d}-symmetric eight-membered crown conformation of S_8 in elemental sulfur is paralleled by molecules such as S_7NH and $S_4N_4H_4$. In the salt $[S_4N_4][AsF_6]_2$, **83**, the dication exists as a planar (D_{4h}) ring that can be classified as a 10π aromatic system,[118] analogous to $[C_8H_8]^{2-}$, its isoelectronic organic counterpart. Likewise, replacement of a cationic sulfur by a methyl-carbon or phenyl-carbon moiety, as in **84**, also yields a planar eight-membered ring structure.[119] In contrast, if the neutral molecule S_4N_4 were to adopt a comparable planar geometry, the 12π electron count would render it antiaromatic; instead, it adopts a D_{2d} cage structure, **85**, in which the four nitrogens are aligned in a square, and pairs of bonded sulfurs are folded above and below this plane.

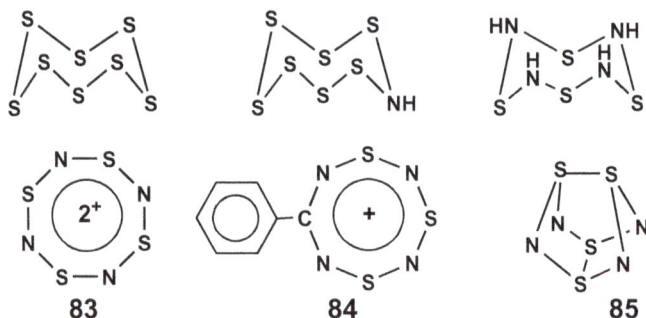

Figure 4.8.   A selection of sulfur-nitrogen molecules and ions.

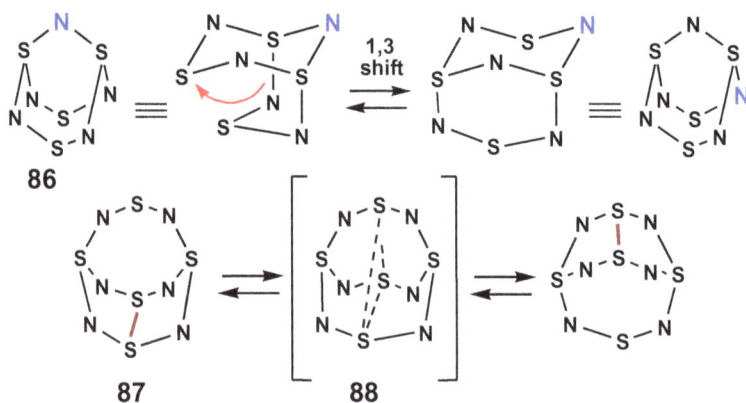

Scheme 4.40.   Gleiter's proposed molecular rearrangements of $[S_4N_5]^+$, **86**, and S_5N_6, **87**.

Derivatives of S_4N_4 are known in which one transannular S–S linkage has been replaced either by a nitrogen atom in $[S_4N_5]^\pm$,[120,121] or by a nitrogen-sulfur-nitrogen bridging moiety in S_5N_6.[122] A molecular orbital study by Bartetzko and Gleiter not only rationalised these structures, but also raised the possibility of their dynamic intramolecular degenerate rearrangement.[123] In particular, it was suggested that in $[S_4N_5]^+$, **86**, a 1,3-shift of an N–S–N bridge from one sulfur to another should be viable, thus allowing the exchange of nitrogens between different environments, as in Scheme 4.40. Furthermore, in S_5N_6, **87**, one could envisage cleavage of the sulfur-sulfur bond to form a transient D_{3h}-symmetric species, **88**, in which all six nitrogens become equivalent;

regeneration of the bond between a different pair of sulfurs brings about a degenerate rearrangement. However, in the S_5N_6 case, a sizeable energy barrier of ~ 30 kcal mol^{-1} was predicted.[123]

Because these are degenerate rearrangements, the detection of any structural changes, and the elucidation of their dynamic behaviour, requires a lowering of their molecular symmetry while maintaining their essential character. In such cases the introduction of isotopically labelled nitrogens at specific sites, thus allowing their detection by ^{15}N NMR spectroscopy, can provide a viable technique to tackle this problem. Indeed, this approach was taken by a world leader in S–N chemistry, Richard Oakley from the University of Waterloo, Ontario, in Canada. The reaction of N,N,N'-tris(trimethylsilyl)benzamide with 99% ^{15}N-enriched $N_3S_3Cl_3$ furnished the required $PhCS_3N_4$, **89**, in which three of the nitrogen sites are labelled, and a cationic sulfur has been replaced by an isoelectronic phenylcarbyne (Ph–C) moiety (Scheme 4.41).[124] (For simplicity and clarity in these rearrangements, to avoid drawing multiple canonical forms, all S–N linkages are shown here as single lines.)

Initially, the ^{15}N NMR spectrum of **89** (Scheme 4.42) exhibits signals at δ 330.1 (N_1, N_2) and at δ 54.9 (N_5) but, over a period of several hours at room temperature, a third signal at δ 176.1 (N_3, N_4) grows in at the expense of the others. This is rationalisable in terms of successive 1,3-migrations of an N–S–N unit from sulfur to carbon, and then of a different N–S–N fragment from carbon back to sulfur; this allows access of the labelled nitrogens to the N_3 and N_4 positions adjacent to carbon.

However, a second rearrangement process is also operative in **89**. The 1,3-migration of the N–C–N unit from sulfur to sulfur (Scheme 4.43) does not bring a ^{15}N-labelled atom into direct contact with the

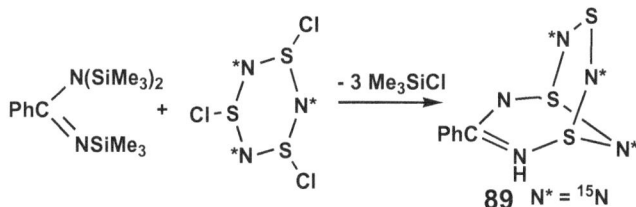

Scheme 4.41.   Synthesis of ^{15}N-labelled $PhCS_3N_5$.

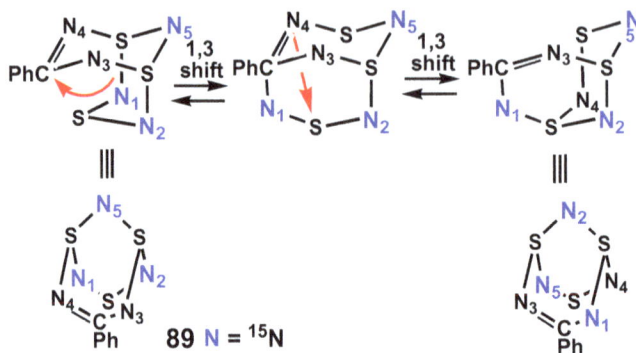

Scheme 4.42. [1,3]-migrations from sulfur to carbon thus equilibrating all nitrogen positions.

Scheme 4.43. [1,3]-migrations from sulfur to sulfur thus equilibrating only the three ^{15}N sites.

carbon, but it does bring about scrambling among the three ^{15}N positions N_1, N_2 and N_5.

A detailed kinetic analysis, including multiple simulations to match the observed experimental data, yielded similar activation energy barriers for the N–S–N migration to and from carbon, and for the second process (N–C–N migration between sulfurs), as 19 ± 1 kcal mol^{-1} and 22 ± 1 kcal mol^{-1}, respectively. As noted by Oakley,[124] "*These pseudodegenerate 1,3-shift reactions suggest that the static picture of molecular structures of sulfur-nitrogen systems, as revealed by X-ray studies, conceals a rich dynamic behaviour.*"

4.8.2. *Isomerisation of 1,3,2,4-dithiadiazolyl Radicals to Disulfides*

The dithianitronium cation, S_2N^+, the sulfur analogue of the nitronium ion, NO_2^+, is a useful building block in the syntheses of more complex sulfur-nitrogen species. Typically, the symmetry-allowed cycloaddition reaction of $[S_2N][AsF_6]$ with nitriles leads to the formation of 6π aromatic 1,3,2,4-dithiadiazolium cations.[125] Moreover, addition of sodium dithionite to a tetrahydrofuran solution of the dithiadiazolium cation brings about reduction to the corresponding dithiadiazolyl radicals, **90**, that are characterised by their EPR spectra. Typically, in the methyl derivative, the spectrum appears as a 1:1:1 triplet with a hyperfine coupling constant a^{N-2} of 1.1 mT, further split by a much smaller coupling, a^{N-4}, of 0.06 mT.

However, when left at room temperature in $SO_2/CFCl_3$ solution for 30 h and longer, the spectrum gradually evolves to yield instead a 1:2:3:2:1 quintet with very small hyperfine coupling, $a^{N-1} = a^{N-4}$ to both nitrogens, clearly indicating rearrangement to the disulfide **91** (Scheme 4.44). This rearrangement of the 1,3,2,4-dithiadiazolyl radical, **90**, to the thermodynamically favoured disulfide **91** appears to proceed via a head-to-tail face-to-face π^*-π^* dimer, **92**. Since calculations reveal that the singly occupied orbital in **90** is a π-type MO based primarily on the sulfur atoms, it is suggested that dimerisation occurs through S–S interactions that subsequently switch partners, as depicted in Scheme 4.45.[126] Although the dimerisation is thermally

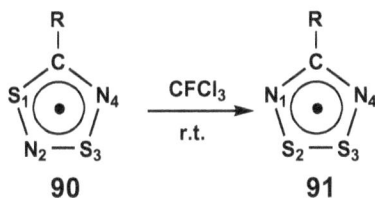

Scheme 4.44. Isomerisation of a dithiadiazolyl radical (R = CH_3, CF_3 or I).

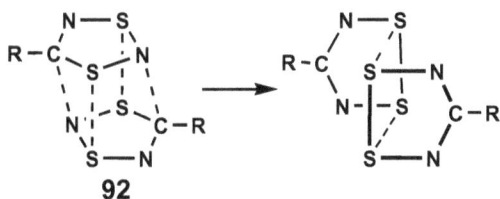

Scheme 4.45. Proposed mechanism for the isomerisation of **90** to **91** via a radical dimer **92**.

symmetry forbidden in the Woodward-Hoffmann sense, it is photo-chemically allowed, and upon photolysis is markedly accelerated.

4.9. Rearrangements of Dicarba-Closo-Dodecaboranes, $C_2B_{10}H_{12}$

The boron hydride anions, $[B_nH_n]^{2-}$, are discussed in more detail in Chapter 7, but we focus here briefly on a product of the reaction of decaborane, $B_{10}H_{14}$, with acetylene to form dicarbadodecaborane, $C_2B_{10}H_{12}$, **93**, with loss of the four B–H–B bridging hydrogens. This reaction is normally performed in the presence of a Lewis base, such as a dialkylsulfide. As shown in Scheme 4.46, the product possesses a closed-cage icosahedral skeletal geometry whereby two adjacent boron atoms in the previously known $[B_{12}H_{12}]^{2-}$ dianion have been replaced by carbons thereby lowering the molecular symmetry from I_h to C_{2v}.[127]

Since their initial preparation in the late 1950s, these molecules, now widely designated as *carboranes*, have undergone a transition from chemical curiosities to important components of thermally stable polymers and, more significantly, have medically important applications, especially against some difficultly treatable cancers. The action of many modern drugs is based around the concept of boron neutron capture therapy (BNCT) whereby boron compounds are selectively concentrated in tumour cells and are then irradiated with low-energy thermal neutrons. The non-radioactive ^{10}B isotope absorbs the neutrons and subsequently breaks up into an α particle and a ^7Li nucleus thus focussing their short-range energy into

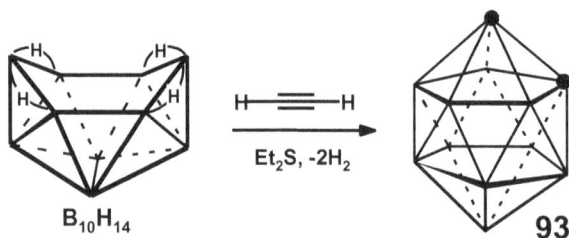

Scheme 4.46.   Formation of *ortho*-carborane $C_2B_{10}H_{12}$ from $B_{10}H_{14}$ and acetylene.

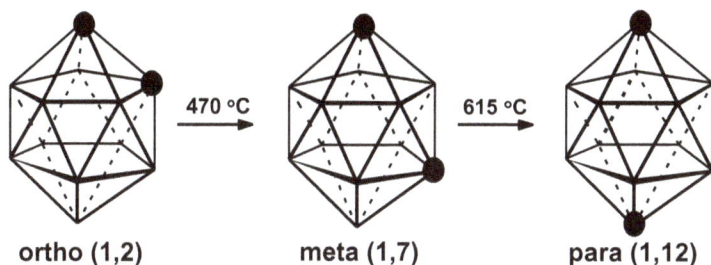

Scheme 4.47. Sequential thermal rearrangement of $C_2B_{10}H_{12}$ isomers, **93** → **94** → **95**.

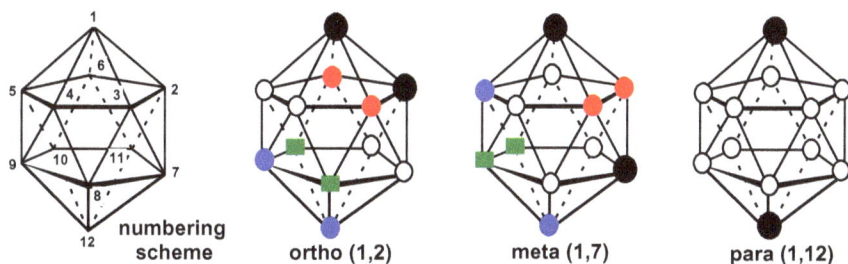

Figure 4.9. The pattern of ^{11}B NMR resonances in the isomers of $C_2B_{10}H_{12}$.

damaging the tumour cell.[128] The advantage of incorporating a carborane moiety into the drug is, of course, its very high boron content.

Of particular interest in our current discussion is the observation that the initially formed product, **93**, in which the carbons occupy adjacent positions in the cage, undergoes rearrangement to the 1,7-isomer, **94**, at temperatures above 470°C;[129] amazingly, this compound is stable up to 615°C,[130] at which point it rearranges into 1,12-$C_2B_{10}H_{12}$, **95** (Scheme 4.47). By analogy to benzene, these isomers are colloquially known as *ortho-*, *meta-* and *para-*carborane.

The three isomers have been characterised by X-ray crystallography, but their structures were initially assigned based on their ^{11}B NMR spectra, originally at relatively low fields,[131] more recently on modern spectrometers by using 2D techniques at much higher frequencies.[132] As indicated in Figure 4.9, the ^{11}B resonances in *ortho*-$C_2B_{10}H_{12}$, **93**, appear as a 2:4:2:2 pattern in accord with its C_{2v} symmetry. Upon conversion to the meta isomer, **94**, the pattern is again 2:4:2:2, once more revealing its C_{2v} symmetry, but the chemical shifts are, of course, slightly different. Finally, in the *para*-isomer, **95**, the molecule is D_{5d}-symmetric and all ten boron nuclei are now equivalent.

The question now arises as to the mechanism of these rearrangements. Despite much discussion, the problem has not been unequivocally resolved; several proposals have been advanced, and it may be the case that more than one mechanism is involved.[133,134] Suggestions have included rotation of a pentagonal pyramid or of a triangular face, but by far the most commonly invoked process is that originally offered by Lipscomb.[135] This involves the formation of an intermediate cuboctahedron, **96**, by opening up triangular faces to form squares. In particular, by cleaving the link between the adjacent carbon atoms and then regenerating the triangles by connecting the other set of diagonally related atoms (in this case, borons) the *ortho* carborane, **93**, is transformed into its *meta* isomer, **94** (Scheme 4.48). This has been termed the *diamond-square-diamond* (*dsd*) process, and eventually, at a sufficiently high temperature, multiple continuations of this procedure could generate the *para* carborane, **95**. However, this final rearrangement may involve rotation of triangular faces, or even mutual rotation of the two pentagonal pyramidal halves of the molecule. These rearrangements are thermodynamically driven

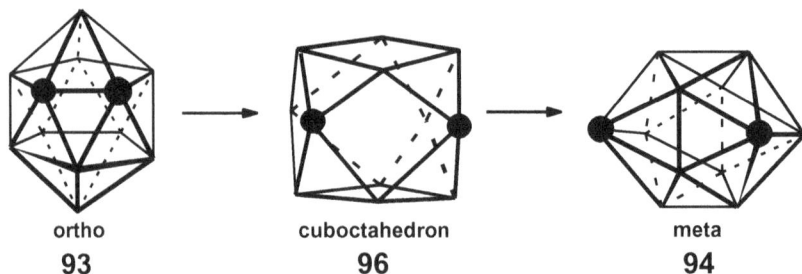

ortho	cuboctahedron	meta
93	**96**	**94**

Scheme 4.48. The *dsd* rearrangement mechanism that converts *ortho* and *meta* carboranes.

by maximising the separation between the relatively positive carbon atoms, culminating in formation of *para*-$C_2B_{10}H_{12}$.

A particularly ingenious experiment was reported by Edvenson and Gaines who prepared specifically ^{10}B-labelled icosahedral carboranes.[136] Since ^{10}B is also an NMR-active nucleus, its location within the cage is readily determined. Their data clearly revealed that, when the 1,2 isomer was thermolysed at temperatures below that required for *ortho* to *meta* rearrangement, the ^{10}B label was completely scrambled. Moreover, mass spectrometric analysis confirmed that these rearrangements were *intra*- rather than *inter*-molecular; that is, there was no intermolecular exchange of labelled boron atoms. In contrast, when the specifically ^{10}B-labelled 1,3 isomer was thermolysed at 470°C (just below the temperature required for *meta* to *para* rearrangement), no isotopic scrambling was observed, nor was there any reversion to the *ortho* isomer. They discussed their data not only in terms of a *dsd* mechanism, but also of triangular face rotations, and even raised the possibility of cage-opening to form a transient *nido* species. More recent high-level calculations at the DFT level suggest that the majority of icosahedral rearrangements proceed via *dsd* steps and triangular face rotations.[134]

4.10. Symmetry with a Twist: Möbius Molecules

4.10.1. *Synthetic Aspects*

As is very well known, taping together the ends of a long thin strip of paper that has been given a single half twist generates a Möbius (Figure 4.10) strip, as depicted in Figure 4.11; it can adopt C_2 symmetry and exist in enantiomeric forms. Such objects have the property that, unlike a ring that has inner and outer surfaces, an insect could walk along, following a single path, and return to the same spot having traversed the entire length of the surface. Moreover, cutting a Möbius strip along the centre line with a pair of scissors yields one long strip with two full twists in it, rather than two separate strips. Even more interesting, when a strip with two half-twists is divided lengthwise it yields a pair of interlinked rings; a similar cutting process

Figure 4.10. August Ferdinand Möbius (1790–1868), mathematician and astronomer.

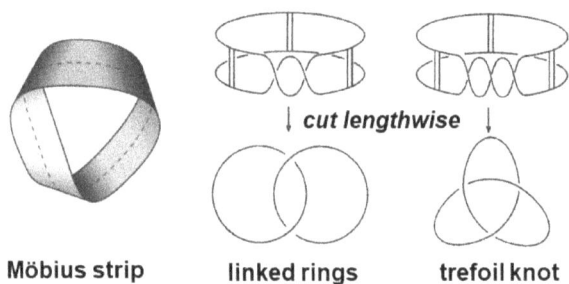

Möbius strip   **linked rings**   **trefoil knot**

Figure 4.11. A Möbius strip and the consequences of lengthwise cutting.

on an object with three half-twists becomes a strip forming a trefoil knot.

Application of these concepts to chemistry was first delineated in a 1961 publication by Frisch and Wasserman from Bell Labs, New Jersey, in which they suggested possible routes to such molecules.[137]

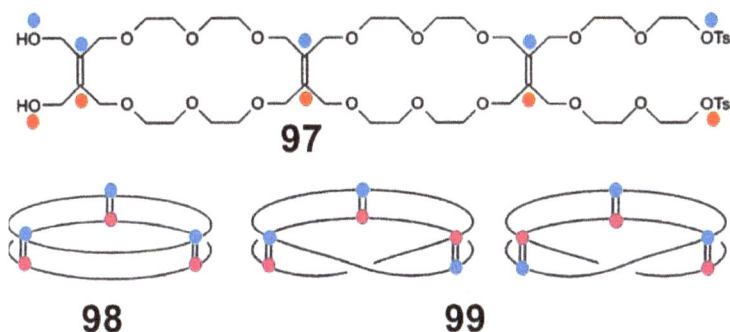

Figure 4.12. Formation of cylindrical and Möbius molecules, **98** and **99**, respectively.

It was first brought to experimental fruition in 1982 as described in a brilliant report by David Walba from the University of Colorado. The approach taken involved the synthesis of a series of linear arrays of crown ethers in which the chains were cross-linked by olefinic double bonds so as to make a molecular ladder, **97**, as depicted in Figure 4.12. In principle, the diol and ditosylate termini of these chains could then be linked either to form a molecular cylinder or a molecular Möbius strip.[138,139] If, upon deprotonation, cyclisation were to occur such that the alkoxides coupled with displacement of the tosylate from the same chains (upper with upper, and lower with lower), then the product would be a cylinder with idealised D_{3h} symmetry. By way of contrast, if coupling were to proceed by cross-linking of upper and lower chains, a Möbius molecule would be formed. Gratifyingly, when the precursor **97** was treated with sodium hydride under rigorously dry conditions (at high dilution to minimise intermolecular interactions), both products were formed and were separable by flash chromatography. The cylindrical D_{3h}-symmetric product, shown schematically as **98**, exhibits a remarkably simple ^{13}C NMR spectrum of only four carbon resonances; it was subsequently further characterised by X-ray crystallography. The isomeric Möbius molecule, **99**, can adopt C_2 as its highest possible symmetry and must therefore be chiral. This latter property was elegantly demonstrated by acquiring its ^{13}C NMR spectrum in a chiral solvent, and the *enantiotopic* olefinic carbons were thereby rendered *diastereotopic*, and therefore NMR non-equivalent; the corresponding spectrum of the cylindrical isomer, **98**, showed no such discrimination.[138]

4.10.2. *Lengthwise Cutting by Ozonolysis*

As noted by Walba, these remarkable Möbius molecules have no chiral centres and are conformationally mobile, yet no molecular rigidity is required to maintain their stereochemical integrity. However, the story does not finish with the mere synthesis and characterisation of the first such molecule. The incorporation of the olefinic linkages now permits the lengthwise cutting procedure, analogous to the use of scissors to cut the paper model. Ozonolysis of the cylindrical molecule, **98**, cleaves the carbon-carbon double bonds that made up the rungs of the molecular ladder to form two molecules of the triketone, **100**. However, the same procedure for the Möbius isomer yields a single large ring, **101**, possessing six ketonic functionalities (Figure 4.13).[140] This was surely a magisterial demonstration of topological chemical control! Since the time of these pioneering experiments, the field of molecular knots and connected rings has grown enormously and has been comprehensively reviewed.[141]

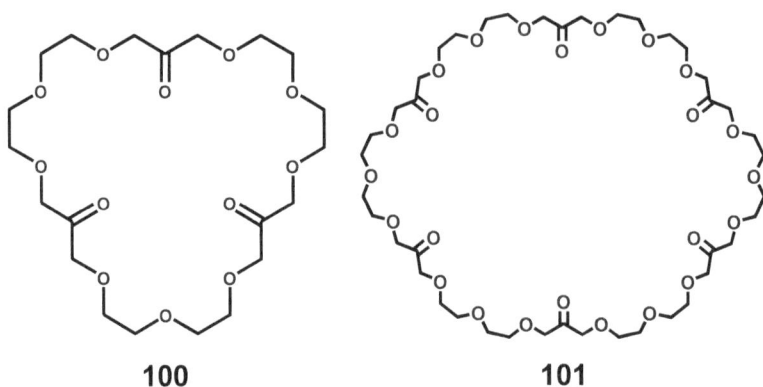

100   **101**

Figure 4.13.   Cyclic ketones **100** and **101** obtained upon ozonolysis of **98** and **99**, respectively.

4.11. Closing Remarks

These examples from across the Periodic Table, selected from the literature published in recent decades, show clearly how symmetry breaking in conjunction with the currently available spectroscopic, spectrometric and other analytical techniques frequently provides an approach whereby mechanistic proposals for reaction mechanisms and molecular rearrangements can be more carefully evaluated. In Chapter 5, we extend this discussion by describing how a wide range of hitherto hidden rearrangements can be detected by judicious lowering of molecular symmetry.

References

1. R.L. Banks and G.C. Bailey, Olefin disproportionation — new catalytic process. *Ind. and Eng. Chem. (Product Res. and Development)* **1964**, *3*, 170.
2. N. Calderon, H.Y. Chen and K.W. Scott, Olefin metathesis — a novel reaction for skeletal transformations of unsaturated hydrocarbons. *Tetrahedron Lett.* **1967**, *34*, 3327–3329.
3. N. Calderon, E.A. Ofstead, J.P. Ward, W.A. Judy and K.W. Scott, Olefin metathesis. I. Acyclic vinylenic hydrocarbons. *J. Am. Chem. Soc.* **1968**, *90*, 4133–4140.
4. C.P. Bradshaw, E.J. Howman and L. Turner, Olefin dismutation: Reactions of olefins on cobalt oxide-molybdenum oxide-alumina. *J. Catal.* **1967**, *7*, 269–276.
5. C.T. Adams and S.G. Brandenberger, Mechanism of olefin disproportionation. *J. Catal.* **1969**, *13*, 360–363.
6. R.H. Grubbs and T.K. Brunk, Possible intermediate in the tungsten-catalyzed olefin metathesis reaction. *J. Am. Chem. Soc.* **1972**, *94*, 2538–2540.
7. G. Lewandos and R. Pettit, Mechanism of the metal-catalyzed disproportionation of olefins. *J. Am. Chem. Soc.* **1971**, *93*, 7087–7088.
8. J.-L. Hérisson and Y. Chauvin, Catalysis of olefin transformations by tungsten complexes. II. Telomerization of cyclic olefins in the presence of acyclic olefins. *Makromol. Chem.* **1971**, *141*, 161–176.
9. Y. Chauvin, Olefin Metathesis: The Early Days (Nobel Lecture). *Angew. Chem. Int. Ed.* **2006**, *45*, 3741–3747.
10. R.R. Schrock, Multiple Metal-Carbon Bonds for Catalysis Reactions (Nobel Lecture). *Angew. Chem. Int. Ed.* **2006**, *45*, 3748–3759.
11. R.H. Grubbs, Olefin Metathesis Catalysts for the Preparation of Molecules and Materials (Nobel Lecture). *Angew. Chem. Int. Ed.* **2006**, *45*, 3760–3765.

12. T.J. Katz and R. Rothchild, Mechanism of the olefin metathesis of 2,2'-divinylbiphenyl. *J. Am. Chem. Soc.* **1976**, *98*, 2519–2526.

13. R.R. Schrock, R.T. DePue, J. Feldman, C.J. Schaverien, J.C. Dewan and A.H. Liu, Preparation and reactivity of several alkylidene complexes of the type W(CHR')(N-2,6-C$_6$H$_3$-*i*-Pr$_2$)(OR)$_2$ and related tungstenacyclobutane complexes. Controlling metathesis activity through the choice of alkoxide ligand. *J. Am. Chem. Soc.* **1988**, *110*, 1423–1435.

14. R.R. Schrock, Olefin metathesis by molybdenum imido alkylidene catalysis. *Tetrahedron* **1999**, *55*, 8141–8153.

15. T.M. Trnka and R.H. Grubbs, The development of L$_2$X$_2$Ru=CHR olefin metathesis catalysts: An organometallic success story. *Acc. Chem. Res.* **2001**, *34*, 18–29.

16. S.T. Nguyen, L.K. Johnson and R.H. Grubbs, Ring-Opening Metathesis Polymerization (ROMP) of norbornene by a Group VIII carbene complex in protic media. *J. Am. Chem. Soc.* **1992**, *114*, 3974–3975.

17. V.P. Kamat, H. Hagiwara, T. Katsumi, T. Susuki and M. Ando, Ring closing metathesis directed synthesis of (*R*)-(-)-muscone from (+)-citronellal. *Tetrahedron* **2000**, *56*, 4397–4403.

18. F. Panella, R.L. Banks and G.C. Bailey, Disproportionation of alkynes. *Chem. Commun.* **1968**, 1548–1549.

19. T. Katz and J. McGinnis, The mechanism of the olefin metathesis reaction. *J. Am. Chem. Soc.* **1975**, *97*, 1592–1594.

20. R.R. Schrock, High-oxidation-state molybdenum and tungsten alkylidyne complexes. *Acc. Chem. Res.* **1986**, *19*, 342–348.

21. M.R. Churchill, J.W. Ziller, J.H. Freudenberger and R.R. Schrock, Metathesis of acetylenes by triphenoxytungstenacyclobutadiene complexes and the crystal structure of W(C$_3$Et$_3$)[O-2,6-C$_6$H$_3$(i-Pr)$_2$]$_3$. *Organometallics* **1984**, *3*, 1554–1562.

22. J. Zhu, G. Jia and Z. Lin, Theoretical investigation of alkyne metathesis catalyzed by W/Mo alkylidene complexes. *Organometallics* **2006**, *25*, 1812–1819.

23. A. Fürstner, Alkyne metathesis on the rise. *Angew. Chem. Int. Ed.* **2013**, *52*, 2794–2819.

24. W. Zhang and J.S. Moore, Shape-persistent macrocycles: Structures and synthetic approaches from arylene and ethynylene building blocks. *Angew. Chem. Int. Ed.* **2006**, *45*, 4416–4439.

25. W. Zhang, S.M. Brombosz, J.L. Mendoza and J.S. Moore, A high yield, one-step synthesis of *o*-phenylene ethynylene cyclic trimer via precipitation-driven alkyne metathesis. *J. Org. Chem.* **2005**, *70*, 10198–10201.

26. T.J. Katz and T.M. Sivavec, Metal-catalyzed rearrangements of alkyne-alkenes and the stereochemistry of metallacyclobutane ring opening. *J. Am. Chem. Soc.* **1985**, *107*, 737–738.

27. S.T. Diver and A.J. Giessert, Enyne metathesis (Enyne bond reorganisation). *Chem. Rev.* **2004**, *104*, 1317–1382.

28. K. Noack, M. Ruch and F. Calderazzo, Carbon monoxide insertion reactions, VI. The mechanisms of the reactions of methylmanganese pentacarbonyl and acetylmanganese pentacarbonyl with triphenylphosphine. *Inorg. Chem.* **1968**, *7*, 345–349.

29. P.M. Maitlis and D. Belli Dell'Amico, Fausto Calderazzo: pioneer in mechanistic organometallic chemistry. *Organometallics* **2014**, *33*, 6989–7006.

30. K. Noack and F. Calderazzo, Carbon monoxide insertion reactions V. The carbonylation of methylmanganesepentacarbonyl with ^{13}CO. *J. Organomet. Chem.* **1967**, *10*, 101–104.

31. J.-A.M. Andersen and J.R. Moss, Synthesis of an extensive series of manganese pentacarbonyl alkyl and acyl compounds: carbonylation and decarbonylation studies on $[Mn(R)(CO)_5]$ and $[Mn(COR)(CO)_5]$. *Organometallics* **1994**, *13*, 5013–5020.

32. T.H. Coffield, J. Kozikowski and R.D. Closson, Acyl manganese pentacarbonyl compounds. *J. Org. Chem.* **1957**, *22*, 598.

33. C.W. Bird, Synthesis of organic compounds by direct carbonylation reactions using metal carbonyls. *Chem. Rev.* **1**, 283–302.

34. T.C. Flood, J.E. Jensen and J.A. Statler, Stereochemistry at manganese of the carbon monoxide insertion in pentacarbonylmethylmanganbese(I). The geometry of the intermediate. *J. Am. Chem. Soc.* **1981**, *103*, 4401–4414.

35. H. Berke and R. Hoffmann, Organometallic migration reactions. *J. Am. Chem. Soc.* **1978**, *100*, 7224–7236.

36. G.M. Whitesides and D.J. Boschetto, Reaction of *threo*-$(CH_3)_2$CHDCHD-Fe(CO)$_2$-π-C$_5$H$_5$ with triphenylphosphine. *J. Am. Chem. Soc.* **1969**, *91*, 4313–4314.

37. D. Seyferth, Chemistry of carbon-function alkylidynetricobalt nonacarbonyl cluster complexes. *Adv. Organomet. Chem.* **1976**, *14*, 97–144.

38. D. Seyferth, J.E. Hallgren and P.L.K. Hung, Organocobalt cluster complexes. 8. Preparation of functional alkylidynetricobalt nonacarbonyl complexes from dicobalt octacarbonyl. *J. Organomet. Chem.* **1973**, *50*, 265–275.

39. R.A. Gates, M.F. D'Agostino, K.A. Sutin, M.J. McGlinchey, T.S. Janik and M.R. Churchill, Syntheses and NMR spectra of Co$_3$(CO)$_9$[μ_3-CCHAr$_2$] clusters derived from DDT and related molecules. X-Ray crystal structures of [[bis(4-chlorophenyl)methyl]-carbynyl]tricobalt nonacarbonyl and of its bis(4-chloronaphthyl) analogue. *Organometallics* **1990**, *9*, 20–26.

40. D. Seyferth, J.E. Hallgren and C.S. Eschbach, Organocobalt cluster complexes. 15. New route to organofunctional organocobalt cluster complexes based on tricobaltcarbon decacarbonyl cation. *J. Am. Chem. Soc.* **1974**, *96*, 1730–1737.

41. H.P. Treffers and L.P. Hammett, Cryoscopic studies on bases in sulfuric acid: the ionization of di-ortho substituted benzoic acids. *J. Am. Chem. Soc.* **1937**, *59*, 1708–1712.

42. D. Seyferth, G.H. Williams and C.L. Nivert, Organocobalt cluster complexes. 15. Conversion of halomethylidynetricobalt nonacarbonyl complexes to tricobaltcarbon decacarbonyl cation by action of aluminum halides — new and useful route to α-functionalized alkylidynetricobalt nonacarbonyl complexes. *Inorg. Chem.* **1977**, *16*, 758–765.

43. M.F. D'Agostino, M. Mlekuz, J.W. Kolis, B.G. Sayer, C.A. Rodger, J.-F-. Halet, J.-Y. Saillard and M.J. McGlinchey, ^{13}C NMR and EHMO studies on Seyferth's $[Co_3(CO)_9CCO]^+$ cluster: to bend or not to bend. *Organometallics* **1986**, *5*, 2345–2350.

44. J.W. Kolis, E.M. Holt and D.W. Shriver, Synthesis, x-ray crystal structure, and chemistry of a metal cluster ketenylidene, $[Fe_3(CO)_9(CCO)]^{2-}$, with carbide-like reactivity. *J. Am. Chem. Soc.* **1983**, *105*, 7307–7313.

45. D. Seyferth and M.O. Nestle, Organocobalt cluster complexes. 23. Novel chemistry of (acylmethylidyne)- and (aroylmethylidyne)tricobalt nonacarbonyl complexes. *J. Am. Chem. Soc.* **1981**, *103*, 3320–3328.

46. R.A. Gates, R.E. Perrier, B.G. Sayer, M.F. D'Agostino and M.J. McGlinchey, The mechanism of decarbonylation of $RC(O)CCo_3(CO)_9$ to $RCCo_3(CO)_9$: Migration of the ketonic carbonyl, *Organometallics* **1987**, *6*, 1181-1187.

47. W.H. Watson, S.G. Bodige, K. Ejsmont, J. Liu and M.G. Richmond, Ligand substitution in $HC(O)Co_3(CO)_9$ with 4,5-bis(diphenylphosphino)-4-cyclopenten-1,3-dione (bpcd): diphosphine ligand fluxionality, decarbonylation of the formyl moiety and competitive P-Ph bond cleavage reactivity. *J. Organomet. Chem.* **2006**, *691*, 3609–3616.

48. R.B. King, Reactions of alkali metal derivatives of metal carbonyls. I. Reactions between $(C_5H_5)Fe(CO)_2Na$ and certain organic polyhalogeno compounds. *Inorg. Chem.* **1963**, *2*, 531–533.

49. R.B. King, Reactions of alkali metal derivatives of metal carbonyls. II. Reactions between acid chloride derivatives and the sodium derivative of cyclopentadienyliron dicarbonyl. *J. Am. Chem. Soc.* **1963**, *85*, 1918–1922.

50. S.C. Kao, C.H. Thiel and R. Pettit, Some aspects of the chemistry of binuclear bridging iron complexes. *Organometallics,* **1983**, *2*, 914–917.

51. J.R. Moss and L.G. Scott, μ(α,ω)-Alkanediyl complexes of transition metals. *Coord. Chem. Rev.* **1984**, *60*, 171–190.

52. R.B. King, Reactions of alkali metal derivatives of metal carbonyls. III. Reaction between sodium pentacarbonylatemanganate (-I) and certain organic polyhalides. *J. Am. Chem. Soc.* **1963**, *85*, 1922–1926.

53. C.P. Casey, The structure of King's $Mn_2(CO)_{10}(CH_2)_3$: a cyclic metal-carbene complex. *Chem. Commun.* **1970**, 1220–1221.

54. C.P. Casey and R.L. Anderson, Metal-carbene complexes from the reaction of methylpentacarbonylmanganese(I) with sodium pentacarbonylmanganate(-I). *J. Am. Chem. Soc.* **1971**, *93*, 3554–3555.

55. J.-A.M. Garner, A. Irving and J.R. Moss, Synthesis of dimanganese nonacarbonyl cyclic carbene complexes and molecular structures of two of them. *Organometallics* **1990**, *9*, 2836–2840.

56. E. Lindner and M. Pabel, 1,3-Propanediylbis(pentacarbonylmanganese) by the bis triflate method. *J. Organomet. Chem.* **1991**, *414*, C19-C21.

57. N.A. Bailey, P.L. Chell, A. Mukhopadhyay, H.E. Tabbron and M.J. Winter, Metal cyclic carbene complexes. A molybdenum complex of 2-oxacyclopentylidene: synthesis and X-ray structure of $[MoI(CO)_2(C_4H_6O)(\eta\text{-}C_5H_5)]$. *J. Chem. Soc. Chem. Commun. 1982*, 215–217.

58. K. Wade, *Electron Deficient Compounds*; Thomas Nelson & Sons: London, UK, 1971.

59. S. Winstein, M. Shatavsky, C. Norton and R.B. Woodward, 7-Norbornenyl and 7-norbornyl cations. *J. Am. Chem. Soc.* **1955**, *77*, 4183–4184.

60. M. Brookhart, R.K. Lustgarten and S. Winstein, Bridge flipping and rearrangement of norbornadienyl and 7-methylnorbornadienyl cations. *J. Am. Chem. Soc.* **1967**, *89*, 6352–6354.

61. S. Winstein and D.S. Trifan, The structure of the bicyclo[2.2.1]2-heptyl (norbornyl) carbonium ion. *J. Am. Chem. Soc.* **1949**, *71*, 2953–2953.

62. S. Winstein and D.S. Trifan, Neighboring carbon and hydrogen. XI. Solvolysis of *exo*-norbornyl *p*-bromobenzenesulfonate. *J. Am. Chem. Soc.* **1952**, *74*, 1154–1160.

63. H.C. Brown, The energy of the transition states and the intermediate cation in the ionization of 2-norbornyl derivatives. Where is the nonclassical stabilization energy? *Acc. Chem. Res.* **1983**, *16*, 432–440.

64. C. Walling, An innocent bystander looks at the 2-norbornyl cation. *Acc. Chem. Res.* **1983**, *16*, 448–454.

65. G.A. Olah, My search for carbocations and their role in Chemistry (Nobel Lecture). *Angew. Chem. Int. Ed. Engl.* **1995**, *34*, 1393–1405.

66. G.A. Olah, K.K. Surya Prakash and M. Saunders. Conclusion of the classical-nonclassical ion controversy based on the structural study of the 2-norborbyl cation. *Acc. Chem. Res.* **1983**, *16*, 440–448.

67. G.A. Olah, G.K. Surya Prakash, M. Arvanaghi and F.A.L. Anet, High-field ¹H and ¹³C NMR spectroscopic study of the 2-norbornyl cation. *J. Am. Chem. Soc.* **1982**, *104*, 7105–7108.

68. P.C. Myhre, K.L. McLaren and C.S. Yannoni, Detection of the static 1,2-dimethyl-2-norbornyl cation by variable-temperature ¹³C CPMAS NMR. *J. Am. Chem. Soc.* **1985**, *107*, 5294–5296.

69. G.A. Olah, A. Commeyras and C.Y. Lui, Stable carbonium ions LXXII. Raman and NMR spectroscopic study of the nortricyclonium ion [protonated

tricyclo[2.2.1.0²,⁶]heptane] and its relation to the 2-norbornyl [bicyclo[2.2.1] heptyl] cation. The nature of the long-lived 2-norbornyl cation in strong acid solution. *J. Am. Chem. Soc.* **1968**, *90*, 3882–3884.

70. G.A. Olah, G.D. Mateescu and J.L. Riemenschneider, Electron spectroscopy of organic ions. II. Carbon 1s electron binding energies of the norbornyl, 2-methylnorbornyl, and related cations. Differentiation between "nonclassical" carbonium and "classical" carbenium ions. *J. Am. Chem. Soc.* **1972**, *94*, 2529–2530.

71. H.C. Brown, The 2-norbornyl cation revisited. *Acc. Chem. Res.* **1986**, *19*, 34.

72. T. Laube, Crystal structure analysis of the 1,2,4,7-*anti*-tetramethyl-2-norbornyl cation: an unsymmetrically bridged carbocation. *Angew. Chem. Int. Ed. Engl.* **1987**, *26*, 560–562.

73. F. Scholz, D. Himmel, F.W. Heinemann, P.v.R. Schleyer, K. Meyer and I. Krossing, Crystal structure determination of the nonclassical 2-bornyl cation. *Science* **2013**, *341*, 62–64.

74. J.B. Thomson, Molecular asymmetry in the ferrocene series. *Tetrahedron Lett.* **1959**, *1(6)*, 26–27.

75. R. Dabard, A. Meyer and G. Jaouen, Preparation of optically active compounds of benchrotrene series. *C.R. Acad. Sci. Ser. C.* **1969**, *268*, 201.

76. R. Dabard and G. Jaouen, Correlations of configurations in the benchrotrene series. *Tetrahedron Lett.* **1969**, *10*, 3391–3394.

77. G. Simonneaux, A. Meyer and G. Jaouen, First examples of resolved enantiomeric chromium(0). *J.C.S. Chem. Comm.* **1975**, 69–70.

78. G. Jaouen and A. Meyer, Some facile syntheses of optically active 2-substituted indanones, indanols, tetralones and tetralols *via* their chromium tricarbonyl complexes. *J. Am. Chem. Soc.* **1975**, *97*, 4667–4672.

79. B. Nichols and M.C. Whiting, The organic chemistry of the transition elements. Part I. Tricarbonylchromium derivatives of aromatic compounds. *J. Chem. Soc.* **1959**, 551–556.

80. L. Tchissambou, R. Dabard and G. Jaouen, Kinetic study on nucleophilic substitution and racemization in benchrotrene series. *C.R. Acad. Sci. Ser. C.* **1972**, *274*, 806.

81. M.F. Semmelhack and H.T. Hall, Intermediates in the reaction of carbanions with π-(chlorobenzene)chromium tricarbonyl. *J. Am. Chem. Soc.* **1974**, *96*, 7092–7093.

82. A. Fretzen, A. Ripa, R. Liu, G. Bernardinelli and E.P. Kündig, 1,2-Disubstituted [η⁶-arene)Cr(CO)₃ complexes by sequential nucleophilic addition/*endo*-hydride abstraction. *Chem. Eur. J.* **1998**, 251–259.

83. F. Rose-Munch and E. Rose, *cine* and *tele* Nucleophilic substitutions in (η⁶-arene)tricarbonylchromium and tricarbonyl(η⁵-cyclohexadienyl)manganese complexes. *Eur. J. Org. Chem.* **2002**, 1269–1283.

84. V. Gagliardini, V. Onnikian, F. Rose-Munch and E. Rose, Chromium hydride intermediates in the case of *cine* and *tele-meta* nucleophilic aromatic substitution on arenetricarbonylchromium complexes. *Inorg. Chim. Acta* **1997**, *259*, 265–271.

85. S.J. Abbott, S.R. Jones, S.A. Weinman and J.R. Knowles, Chiral [^{16}O,^{17}O,^{18}O] phosphate monoesters. 1. Asymmetric synthesis and stereochemical analysis of [1(R)-^{16}O,^{17}O,^{18}O]phospho-(S)-propane-1,2-diol. *J. Am. Chem. Soc.* **1978**, *100*, 2558–2560.

86. S.L. Buchwald and J.R. Knowles, Determination of the absolute configuration of [^{16}O,^{17}O,^{18}O]phosphate monoesters by using ^{31}P NMR. *J. Am. Chem. Soc.* **1980**, *102*, 6601–6602.

87. S.R. Jones, L.A. Kindman and J.R. Knowles, Stereochemistry of phosphoryl group transfer using a chiral [^{16}O,^{17}O,^{18}O] stereochemical course of alkaline phosphatase. *Nature* **1978**, *275*, 564–565.

88. G. Lowe, Chiral [^{16}O,^{17}O,^{18}O]phosphate esters. *Acc. Chem. Res.* **1983**, *16*, 244–251.

89. J. Lüthy, J. Rétey and D. Arigoni, Preparation and detection of chiral methyl groups. *Nature* **1969**, *221*, 1213–1215.

90. J.W. Cornforth, J.W. Redmond, H. Eggerer, W. Buckel and C. Gutschow, Asymmetric methyl groups. *Nature* **1969**, *221*, 1212–1213.

91. J.W. Cornforth, J.W. Redmond, H. Eggerer, W. Buckel and C. Gutschow, Synthesis and configurational assay of asymmetric methyl groups. *Eur. J. Biochem.* **1970**, *14*, 1–13.

92. C.A. Townsend, T. Scholl and D. Arigoni, A new synthesis of chiral acetic acid. *J.C.S. Chem. Comm.* **1975**, 921–922.

93. K. Kobayashi, P.K. Jadhav, T.M. Zydowsky and H.G. Floss, A simple and efficient synthesis of chiral acetic acid of high optical purity. *J. Org. Chem.* **1983**, *48*, 3510–3512.

94. L.J. Altman, C.Y. Han, A. Bertolino, G. Handy, D. Laungani, W. Muller, S. Schwartz, D. Shanker, W.H. de Wolf and F. Yang, Stereochemistry of the 1,3-proton loss from a chiral methyl group in the biosynthesis of cycloartenol as determined by tritium nuclear magnetic resonance spectroscopy. *J. Am. Chem. Soc.* **1978**, *100*, 3235–3237.

95. H.G. Floss and S. Lee, Chiral methyl groups: small is beautiful. *Acc. Chem. Res.* **1993**, *26*, 116–122.

96. L.D. Zydowsky, T.M. Zydowski, E.S. Haas, J.W. Brown, J.W. Reeve and H.G. Floss. Stereochemical course of methyl transfer from methanol to methyl coenzyme M in cell-free extracts of *Methanosarcina barkeri*. *J. Am. Chem. Soc.* **1987**, *109*, 7922–7923.

97. R.A. Marcus, Electron transfer reactions in chemistry. Theory and practice. (Nobel lecture). *Angew. Chem. Int. Ed.* **1993**, *32*, 1111–1121.

98. N.S. Hush, Distance dependence of electron-transfer rates. *Coord. Chem. Rev.* **1985**, *64*, 135–157.

99. M.-S. Chan and A.C. Wahl, Rate of electron exchange between iron, ruthenium and osmium complexes containing 1,10 phenanthroline, 2,2'-bipyridyl, or their derivatives from nuclear magnetic resonance studies. *J. Phys. Chem.* **1978**, *82*, 2542–2549.

100. R.J. Campion, C.F. Deck, P. King, Jr. and A.C. Wahl, Kinetics of electron exchange between hexacyanoferrate(II) and (III) ions. *Inorg. Chem.* **1967**, *6*, 672–681.

101. R.J. Prestwood and A.C. Wahl, Kinetics of the thallium(I)–thallium(III) exchange reaction. *J. Am. Chem. Soc.* **1949**, *71*, 3137–3145.

102. J.C. Sheppard and A.C. Wahl, Rate of electron-transfer exchange between manganate and permanganate ions. *J. Am. Chem. Soc.* **1953**, *75*, 5133–5134.

103. J.C. Sheppard and A.C. Wahl, Kinetics of the manganate–permanganate exchange reaction. *J. Am. Chem. Soc.* **1957**, *79*, 1020–1024.

104. K.V. Krishnamurty and A.C. Wahl, Kinetics of the vanadium(II)–vanadium(III) exchange reaction. *J. Am. Chem. Soc.* **1958**, *80*, 5921–5924.

105. B.M. Gordon and A.C. Wahl, Kinetics of the silver(I)–silver(II) exchange reaction. *J. Am. Chem. Soc.* **1958**, *80*, 273–276.

106. C.H. Cheek, N.A. Bonner and A.C. Wahl, Antimony (III)–antimony (V) exchange in HCl solutions. *J. Am. Chem. Soc.* **1961**, *83*, 80–84.

107. V.J. Linnenbom and A.C. Wahl, Exchange reactions between cerium(III) and cerium(IV) and between iron(II) and iron(III). *J. Am. Chem. Soc.* **1949**, *71*, 2589–2590.

108. A. Werner, Concerning mirror image isomerism in iron compounds. *Ber.* **1912**, *45*, 433–436.

109. F.P. Dwyer and E.C. Gyarfas, The preparation of the optical forms of tris-2,2'-dipyridyl iron (III) and tris-1,10-phenanthroline osmium(III) perchlorates. *J. Am. Chem. Soc.* **1952**, *74*, 4699–4700.

110. F.H. Burstall, F.P. Dwyer and E.C. Gyarfas, Optical activity dependent on a six-coordinate bivalent osmium complex. *J. Chem. Soc.* **1950**, 953–955.

111. F.P. Dwyer and E.C. Gyarfas, A reaction for the study of the kinetics of electron transfer. *Nature* **1950**, *166*, 481.

112. H. Taube, Electron transfer between metal complexes — a retrospective view (Nobel Lecture). *Angew. Chem. Int. Ed. Engl.* **1984**, *23*, 329–394.

113. F. Paul and C. Lapinte, Organometallic molecular wires and other nanoscale-sized devices. An approach using the organoiron (dppe)Cp*Fe building block. *Coord. Chem. Rev.* **1998**, *178–180*, 431–509.

114. A. Wong, P.C.W. Kang, C.D. Tagge and D.R. Leon, Synthesis and characterization of bimetallic complexes with the bridging $\eta^2(\sigma,\sigma)$-1,3-butadiyl ligand. *Organometallics* **1990**, *9*, 1992–1994.

115. T. Bartik, B. Bartik, M. Brady, R. Dembinskii and J.A. Gladysz, A step-growth approach to metal-capped one-dimensional carbon alloptropes: syntheses of C_{12}, C_{16}, and C_{20} μ-polyynediyl complexes. *Angew. Chem. Int. Ed. Engl.* **1996**, *35*, 414–417.

116. A.M. Stolzenberg and E.L. Muetterties, Mechanisms of $Re_2(CO)_{10}$ substitution reactions: crossover experiments with $^{185}Re_2(CO)_{10}$ and $^{187}Re_2(CO)_{10}$. *J. Am. Chem. Soc.* **1983**, *105*, 822–827.

117. T. Chivers, *A Guide to Chalcogen-Nitrogen Chemistry*, World Scientific: Singapore, 2005.

118. R.J. Gillespie, J.P. Kent, J.F. Sawyer, D.T. Slim and J.D. Tyrer, Reactions of sulfur nitride (S_4N_4) with antimony pentachloride, antimony pentafluoride, and fluorosulfuric acid. Preparation and crystal structures of salts of the $S_4N_4^{2+}$ cation: $(S_4N_4)(Sb_3F_{14})$ (SbF_6), $(S_4N_4)(SO_3F)_2$, $(S4N4)(AsF_6)_2 \cdot SO_2$, (S_4N_4) $(AlCl_4)_2$, and $(S_4N_4)(SbCl_6)_2$. *Inorg. Chem.* **1981**, *20*, 3799–3812.

119. C. Knapp, P.G. Watson, E. Lork, D.H. Friese, R. Mews and A. Decken, Trithiatetrazocine cations, $[RCN_4S_3]^+$; planar sulfur-nitrogen 10π aromatics. *Inorg. Chem.* **2008**, *47*, 10618–10625.

120. W. Flues, O.J. Scherer, J. Weiss and G. Wollmershäuser, Crystal and molecular structure of the tetrasulfur pentanitride anion. *Angew. Chem. Int. Ed. Engl.* **1976**, *15*, 379–380.

121. T. Chivers and L. Fielding, A bicyclic S-N cation: the synthesis and crystal structure of S_4N_5Cl. *J. Chem. Soc. Chem. Commun.* **1978**, 212–213.

122. T. Chivers and J. Proctor, Preparation and crystal structure of a new sulfur nitride, S_5N_6; a molecular basket. *J. Chem. Soc. Chem. Commun.* **1978**, 642–643.

123. R. Bartetzko and R. Gleiter, The structures of S_4N5^-, $S_4N_5^+$ and S_5N_6. A rationalization based upon molecular orbital theory. *Chem. Ber.* **1980**, *113*, 1138–1144.

124. K.T. Bestari, R. Boeré and R.T. Oakley, Degenerate and pseudodegenerate 1,3-nitrogen shifts in sulfur-nitrogen chemistry. ^{15}N NMR analysis of skeletal scrambling in $PhCN_5S_3$. *J. Am. Chem. Soc.* **1989**, *111*, 1579–1584.

125. G.K. MacLean, J. Passmore, M.N. Sudneedra Rao, M.J. Schriver, P.S. White, D. Bethell, R.S. Pilkington and L.H. Sutcliffe, Preparation of 1,3,2-dithiazolium hexafluoroarsenate(V), preparation and crystal structures of 5-methyl-1,3,2,4-dithiadiazolium hexafluoroarsenate and 4-methyl-1,3,2-dithiazolium hexafluoroarsenate(V) and the reduction of these salts to stable free radicals. *J. Chem. Soc. Dalton Trans.* **1985**, 1405–1416.

126. N. Burford, J. Passmore and M.J. Schriver, The preparation and isolation of 5-methyl-1,3,2,4-dithiadiazolyl and the facile rearrangement of the 1,3,2,4-dithiadiazolyl radicals to the disulfide isomers 2,3,1,4-dithiadiazolyl. *J. Chem. Soc. Chem. Commun.* **1986**, 140–142.

127. T.L. Heying, J.W. Ager, Jr., S.L. Clark, D.J. Mangold, H.L. Goldstein, M. Hillman, R.J. Polak and J.W. Szymanski, A new series of organoboranes. I. Carboranes from the reaction of decaborane with acetylenic compounds. *Inorg. Chem.* **1963**, *2*, 1089–1092.

128. K. Nedunchezhian, N. Aswath, M. Thiruppathy and S. Thirugnanmurthy, Boron neutron capture therapy — a literature review. *J. Clin. Diagn. Res.* **2016**, *10*, Ze01-ZE04.

129. D. Grafstein and J. Dvorak, Neocarboranes, a new family of stable organoboranes isomeric with the carboranes. *Inorg. Chem.* **1963**, *2*, 1128–1133.

130. S. Papetti and T.L. Heying, *p*-Carborane [1,12-dicarbclovododecaborane]. *J. Am. Chem. Soc.* **1964**, *86*, 2295.

131. G.D. Vickers, H. Agahigian, E.A. Pier and H. Schroeder, Elucidation of boron (^{11}B) nuclear magnetic resonance spectra by heteronuclear spin coupling. *Inorg. Chem.* **1966**, *5*, 1089–1092.

132. S. Heřmánek, ^{11}B NMR spectra of boranes, main-group heteroboranes, and substituted derivatives. Factors influencing chemical shifts of skeletal atoms. *Chem. Rev.* **1992**, *92*, 325–362.

133. Y.V. Roberts and B.F.G. Johnson, Dicarbadodecaborane rearrangements: an appraisal of rotational mechanisms. *J. Chem. Soc. Dalton Trans.* **1994**, 759–766.

134. C.A. Brown and M.L. McKee, Rearrangements in icosahedral boranes and carboranes revisited. *J. Mol. Model.* **2006**, *12*, 653–664.

135. A. Kaczmarcyck, R.D. Dobrott and W.N. Lipscomb, Reactions of $[B_{10}H_{10}]^{2-}$ ion. *Proc. Natl. Acad. Sci. USA* **1962**, *48*, 729.

136. G.M. Edvenson and D.F. Gaines, Thermal isomerization of regiospecifically ^{10}B-labeled icosahedral carboranes. *Inorg. Chem.* **1990**, *29*, 1210–1216.

137. H.L. Frisch and E. Wasserman, Chemical topology. *J. Am. Chem. Soc.* **1961**, *83*, 3789–3795.

138. D.M. Walba, R.M. Richards and R.C. Haltowanger, Total synthesis of the first molecular Möbius strip. *J. Am. Chem. Soc.* **1982**, *104*, 3219–3221.

139. D.M. Walba, Topological stereochemistry. *Tetrahedron* **1985**, *41*, 3161–3212.

140. R. Herges, Topology in chemistry: designing Möbius molecules. *Chem. Rev.* **2006**, *106*, 4820–4842.

141. R.S. Forgan, J-P. Sauvage and J.F. Stoddart, Chemical topology: complex molecular knots, links and entanglements. *Chem. Rev.* **2011**, *111*, 5434–5464.

Chapter 5

The Detection and Elucidation of Hidden Molecular Rearrangements

"Three things cannot long be hidden: the sun, the moon and the truth."

— *Confucius*

5.1. The Significance of Symmetry Breaking in NMR Spectroscopy

In previous chapters we have discussed a wide range of molecular rearrangements, from those in which a starting material goes irreversibly to product, as in the Curtius or Beckmann, to self-exchange redox processes such as the $[Fe(CN)_6]^{4-}/[Fe(CN)_6]^{3-}$ system where the reactants and products are identical. In the latter cases, symmetry breaking by taking advantage of chirality changes or the incorporation of radioactive isotopes has been utilised. We focus here on the applications of variable-temperature NMR spectroscopy that is probably the most convenient and sensitive technique to monitor changes in molecular structure in solution.

Symmetry breaking plays a crucial role in many aspects of NMR spectroscopy. For example, theoreticians calculating the simplest spin-spin coupling constant, $^1J_{HH}$ in dihydrogen, need an experimental

121

measurement to validate their predictions.[1] Evidently, this is not obtainable from H_2 itself since the two nuclei are equivalent and the observed gas phase spectrum is a singlet. However, isotopic substitution, as in HD, yields duplicate values of $^1J_{HD}$ as 43.3 Hz, not only from the 1:1:1 triplet in the proton spectrum, but also from the 1:1 doublet in the deuterium spectrum (the nuclear spin values, I, for ^1H and ^2D are ½ and 1, respectively). The unobservable $^1J_{HH}$ is now readily calculated since $^1J_{HH}/^1J_{HD} = \gamma_H/\gamma_D = 6.51$, where γ is the magnetogyric ratio for the relevant nucleus; the experimental value for $^1J_{HH}$ is therefore 282 Hz.

It is commonly the case that molecular rearrangements occur very rapidly at room temperature on the NMR timescale. They exhibit simplified spectra whereby non-equivalent nuclear environments yield time-averaged resonances, thus equilibrating nuclear environments that are in fact non-equivalent in the static system, as seen for example by X-ray crystallography. At lower temperatures, when the rate of exchange is sufficiently reduced, these degeneracies are split and the underlying "static" molecular symmetry becomes apparent. Variable-temperature NMR spectroscopy frequently allows elucidation not only of the mechanism of rearrangement, but also the activation energies and entropies of the process or processes involved. Frequently, however, such rearrangement processes are hidden, even when they become slow on the NMR timescale, because the molecular point group remains unchanged. In many cases, this situation can be resolved by judicious symmetry breaking, such as substitution of a molecular fragment by a similar, but not identical, moiety or the incorporation of potentially diastereotopic nuclei, thus allowing the elucidation of the kinetics and energetics of such processes. Examples are chosen that include a wide range of rotations, migrations and other rearrangements in organic, inorganic and organometallic chemistry.

As noted in Chapter 3, bullvalene, a $C_{10}H_{10}$ isomer, undergoes rapid multiple Cope rearrangements and exhibits a single resonance in both the ^1H and ^{13}C NMR regimes at 120°C, but at low temperatures each is split into a 3:3:3:1 peak ratio, in accord with the solid-state structure revealed by X-ray crystallography. However, there is no need to introduce additional labels to break the three-fold symmetry since it is immediately exposed merely by lowering the temperature; this is not what we regard as a hidden rearrangement in terms of the present discussion. Instead, as we have already discussed in Chapters 3 and 4, the 1,5-hydrogen shifts in heptaarylcyclohepta-trienes, or Oakley's demonstration of the dynamic behaviour of sulfur-nitrogen cage compounds, are only revealed by the incorporation of a symmetry-breaking entity.

5.2. The Role of Diastereotopic Nuclei in Dynamic Processes

At this point it may be useful to remind ourselves of the differences between homotopic, enantiotopic and diastereotopic nuclei in NMR spectra. Considering for the moment only ^{1}H spectra, in each case it is instructive to replace one of the nuclei under investigation by deuterium and then classify the resulting product. Starting from ethane, **1**, let us separately replace each of the two labelled protons by deuterium, as in Scheme 5.1. It is clear that the product is the same in each case, and so these nuclei are *homotopic*.

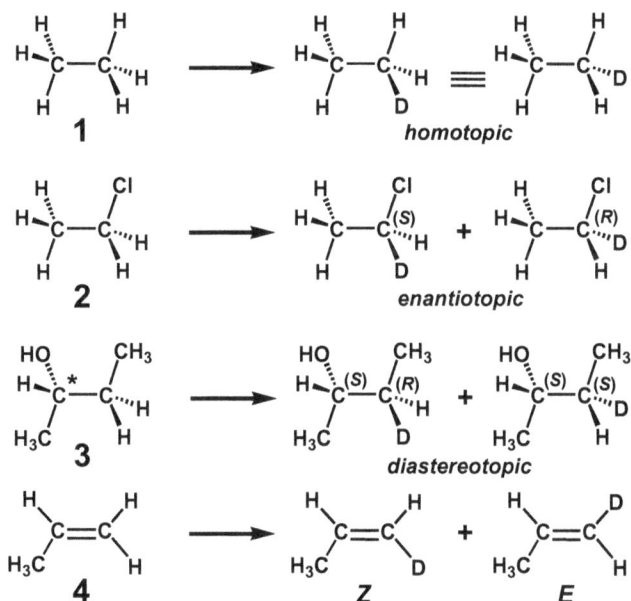

Scheme 5.1. Examples of homotopic, enantiotopic and diastereotopic nuclei.

Now, beginning from chloroethane, **2**, let us replace separately each of the methylene protons by deuterium and note that the products are enantiomers; these nuclei are *enantiotopic*. They still resonate at the same frequency and are indistinguishable by NMR; we note, however, that enzymes readily differentiate between such proton pairs, thus generating single enantiomers in biological processes.

Finally, we commence with a molecule such as (*S*)-butan-2-ol, **3**, bearing a single stereocentre and once again substitute a deuterium for a hydrogen. It is evident that the products are diastereomers and that the protons under consideration are *diastereotopic*. Not only are these nuclei differentiable by NMR, but also they couple to each other thereby giving rise to a pair of doublets further split by the single proton at C(2). *Diastereotopism* also arises by replacement of a geminal hydrogen, for example by deuterium in propene, **4**, to form either an *E* or *Z* product.

In very early reports, the non-equivalence of the geminal fluorines in $PhCHBr-CF_2Br$ was originally explained in terms of restricted rotation about the central C–C bond,[2] and it was only later accepted that this phenomenon was an indication of the presence of a neighbouring stereocentre, as delineated in a seminal review by Mislow and Raban.[3] Its realisation in the ^{13}C regime was reported by Sokolov who prepared an α-ferrocenyl carbocation, **5**, bearing an isopropyl substituent.[4] Knowing the high stability of such cations as the result of a direct interaction of the cationic carbon with the metal centre,[5,6] he surmised correctly that the isopropyl methyls should be diastereotopic and so give rise to separate ^{13}C resonances. Moreover, it was noted that in the diferrocenylmethyl cation, **6**, the C_5H_4 ring protons only exhibit diastereotopism at low temperature when rotation about the cyclopentadienyl-C(H) linkages become slow on the NMR timescale (Scheme 5.2).[5]

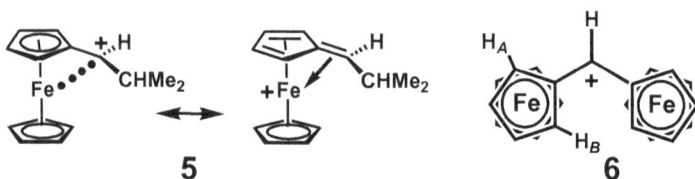

Scheme 5.2.   Diastereotopism in ferrocenyl carbocations.

Diastereotopic isopropyl methyls as probes for chirality have since been very widely used not only for structural characterisation, but also to monitor dynamic behaviour. We note in particular the approach used by Hackett Bushweller, then at Worcester Polytech in Massachussetts, subsequently at the University of Vermont, who recorded the variable-temperature ^{1}H NMR spectra of a series of *N*-isopropyl-*N*,*N*-dialkylamines.[7] Typically, in $CH_3(C_2D_5)NCH(CH_3)_2$, 7, at ambient temperature the isopropyl methyls appeared as a single resonance, doublet split by the adjacent hydrogen. However, at the very low temperature of 130 K, decoalescence of this signal revealed a pair of doublets showing that the methyl groups had become non-equivalent attributable to slowed nitrogen inversion on the NMR timescale. In this chiral system, the isopropyl methyl groups are intrinsically diastereotopic and only become time-averaged as the molecule inverts, in an umbrella-like manner, via a mirror-symmetric transition state (Scheme 5.3). Simulation of the spectra yielded a value of 7.4 ± 0.2 kcal mol^{-1} for the inversion barrier. Since then, this work has been extended into the ^{13}C NMR regime and, when combined with molecular mechanics calculations, also provided barriers for rotation about the alkyl-nitrogen linkages.[8]

In contrast, since tertiary phosphines of the type $PR^1R^2R^3$ can be separated into stable enantiomers, the inversion barrier must be substantially higher than the corresponding process in amines. This was nicely shown by Fluck and Isslieb who noted that 1,2-dimethyl-1,2-diphenyldiphosphine PhMeP–PMePh, 8, is found in two isomeric forms, *meso*-8 and *d,l*-8, and gives rise to two ^{31}P NMR resonances.[9] This system was also studied by Lambert and Mueller who monitored

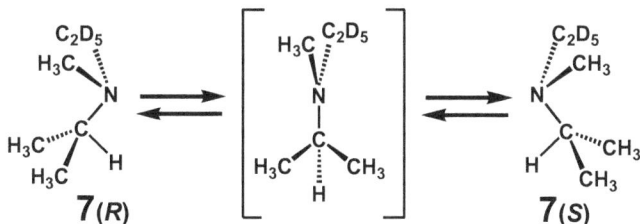

Scheme 5.3. Inversion at nitrogen in an isopropyl-substituted amine.

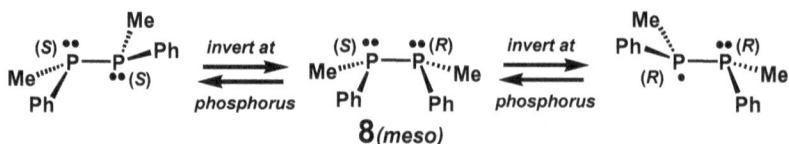

Scheme 5.4. Inversion at phosphorus leads to interconversion of *meso-* and *d,l-*diphosphines.

the variable-temperature ^{1}H NMR behaviour of **8**.[10] As shown in Scheme 5.4, inversion at a single phosphorus center brings about interconversion of these diastereomers which exhibit two methyl signals at room temperature but start to coalesce above 130°C. The activation energy for inversion at phosphorus was measured as 26 ± 2 kcal mol^{-1}; the inversion barrier in the corresponding diarsine was 27 ± 1 kcal mol^{-1}.[11] At the time, these high barriers were thought to be a consequence of pπ-dπ bonding, but nowadays this is not considered to be a viable explanation.[12] Interestingly, the barrier to inversion at another second row element, sulfur in chiral sulfoxides, lies in the range 35–42 kcal mol^{-1}.[13]

Taking their cue from a suggestion by Pirkle,[14] in an extremely ingenious experiment, Albrand and Robert, from Grenoble in France, devised a method to identify which of the ^{31}P resonances in 1,2-diphenyldiphosphine PhPH–PHPh, **9**, can be assigned to *meso-***9**, and which to the *d,l* diastereomer.[15] In a regular achiral solvent (pyridine) the proton-decoupled ^{31}P NMR spectrum appears as two singlets at 67.6 and 71.2 ppm. However, when recorded in a chiral solvent, the two phosphorus nuclei in the *meso* isomer (which are normally enantiotopic, one *R*, the other *S*) now become diastereotopic (Scheme 5.5), and should couple to each other, thus giving rise to an AB pattern. In contrast, while the two ^{31}P nuclei *within* each individual *R,R* or *S,S* isomer retain their equivalence, these enantiomers are themselves diastereotopic in the chiral environment, and should appear as a pair of singlets.

Accordingly, when the ^{31}P spectrum was acquired in a chiral solvent, (+)-1-phenylethylamine, at 40.5 MHz (on a 100 MHz spectrometer), the resonance at 71.2 ppm is clearly split into two peaks, but it is necessary to determine whether this is a chemical shift

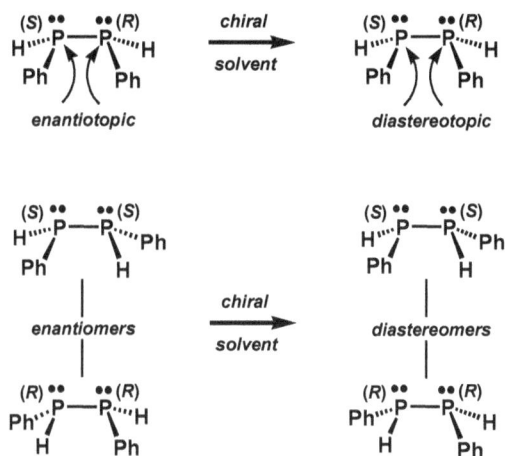

Scheme 5.5. The behaviour of *meso-* and *d,l*-1,2-diphenyldiphosphine, **9**, in a chiral solvent.

difference or part of an AB coupling pattern with very weak outer lines. This was resolved by acquiring the spectrum again at a higher field on a 270 MHz spectrometer whereby, in the case of a chemical shift difference, the separation *in ppm* should be unchanged, but *in Hz* should increase by a factor of 2.7 (the ratio of the spectrometer frequencies), and this was in fact observed for the resonance at 71.2 ppm now assignable to *d,l*-PhPH-PHPh. The AB pattern for the *meso* resonance at 67.6 ppm is also slightly modified, but the coupling constants are unchanged.

We note that diastereotopism has been observed for a number of other nuclei; there are, of course, very many examples for ^{13}C, ^{19}F, and ^{31}P, but they have also been reported for other nuclei such as ^2H, ^3H, ^7Li, ^{15}N, and ^{17}O.[16-20]

5.3. The Measurement of Activation Energies in Exchange Processes

Let us briefly remind ourselves of some of the most widely used approaches for extracting activation energies for molecular rearrangements from variable-temperature NMR data. In a system in which a number of resonances seen at low temperature gradually coalesce upon heating to yield a simpler spectrum, we need to know the magnitude of the rate constant, k, for each experimental spectrum at a given temperature, T. Then, knowing the values for k and T over a suitably wide temperature range, an Arrhenius plot of $\ln k$ vs $1/T$ yields $\Delta G^{\ddagger}$ for the process.

To obtain a value for k for each experimental temperature T, we can simulate the spectrum by using a programme[21,22] into which we input the chemical shifts, coupling constants and peak intensities from the low-temperature limiting spectrum, together with information as to which particular resonances we believe undergo exchange. We then assign a list of rate constants and run the programme to generate the predicted spectrum for each value of k. This process is then refined by modifying the values of k so as to find the closest possible fit of the simulated spectrum to the experimental one for each temperature, thereby yielding the values of k and T required for the Arrhenius plot to obtain $\Delta G^{\ddagger}$. This procedure is relatively straightforward and reliable, but can be rather time-consuming as multiple simulations are normally required to obtain the best match between experimental and simulated spectra. If we have very good data, acquired over a wide temperature range, then a plot of $\ln k/T$ vs $1/T$ in an Eyring plot yields values of $\Delta H^{\ddagger}$ and $\Delta S^{\ddagger}$. The need for particularly reliable data arises because even a small change in the slope of the Eyring plot that yields $\Delta H^{\ddagger}$ can lead to a much larger change (and potential error) in the y intercept from which the activation entropy, $\Delta S^{\ddagger}$, is obtained.

A more convenient approach is to simplify the system such that the behaviour of only a small number of resonances, preferably two, need to be monitored over the required temperature range. As explicated by Gutowsky and Holm,[23] the situation can be reduced to a single equation. Consider the simplest case of two equal-intensity singlets (no coupling), at frequencies v_A and v_B. As the temperature increases, they will gradually collapse, passing via a broad, almost flat-top, peak (at the coalescence temperature, T_C), eventually becoming a single sharp resonance. At T_C, where the two separate peaks have just coalesced, the lifetime, τ, of nucleus A (or B) is given by Equation (5.1):

$$\tau = \sqrt{2}/(\pi \, \delta v), \quad \text{where } \delta v = v_A - v_B \text{ (in Hz).} \quad (5.1)$$

The rate constant for the reaction A → B is the inverse of the lifetime,

$$k = 1/\tau,$$

but we know that

$$k = (RT/Nh)e^{-\Delta G\ddagger/RT},$$

therefore at coalescence,

$$\Delta G\ddagger/RT_C = \ln(\sqrt{2}R/\pi Nh) + \ln(T_C/\delta v)$$

where R is taken as 1.987 cal deg^{-1} mol^{-1}, thus

$$\Delta G\ddagger/RT_C = 22.96 + \ln(T_C/\delta v). \tag{5.2}$$

Experimentally, we only need to know the separation between v_A and v_B, in Hz, and the coalescence temperature, in K, to evaluate the activation energy; Equation (5.2) gives the value of $\Delta G\ddagger$ in calories. Of course, more sophisticated approaches including 2D-exchange methods are now routinely used and have greatly extended the scope of these endeavours.[24]

The advantages of incorporating diastereotopic nuclei as labels to monitor the progress of an exchange process are now apparent. In particular, the ^{13}C NMR spectrum of an isopropyl group, in cases where the methyls are diastereotopic, appears as a pair of equal-intensity singlets in the low-temperature limit, but can pass via an easily recognisable coalescence temperature *en route* to an "exchange-averaged" singlet. We shall illustrate this approach with a number of chemically very different examples.

We should perhaps dispel a concern frequently raised in questions after seminars on the study of rearrangements by NMR methods. It is not necessary to work with optically pure reactants since the spectra of enantiomers are identical, unless a chiral solvent is used (as was the case when distinguishing between *meso-* and *d,l*-diphosphines discussed in Section 5.2.) The diastereotopic nuclei under investigation in both enantiomers appear as separate resonances until the increasing rate of exchange brings about peak coalescence.

5.4. Isopropyl Groups as Mechanistic Probes for Rearrangement Processes

5.4.1. *Tris-chelate Complexes M(L–L)$_3$*

Complexes such as tri(acetylacetato)chromium(III) or the tris(ethylenediamine)cobalt(III) cation possess D_3 symmetry and are chiral. As shown in the classic work by Werner, discussed in Chapter 1, when the latter compound is partnered by a chiral anion, e.g. (+)-tartrate, the diastereomers can be separated by fractional crystallisation, and the enantiomeric cations can each be obtained subsequently in optically pure form. However, in some cases, systems of this general type racemise more rapidly than can be conveniently monitored by polarimetry or other related techniques. Nevertheless, such processes can be readily followed by NMR spectroscopy by incorporating a potentially diastereotopic substituent to probe the chirality. As noted above, it is not necessary to work with optically pure compounds since both enantiomers exhibit the same spectrum.

An important early contribution by Jurado and Springer involved the dynamic behaviour of tri(acetylacetato)aluminium, which adopts D_3 symmetry. Incorporation of isopropyl groups, shown schematically in **10**, lowers the molecular symmetry to C_3 and provides diastereotopic methyls to monitor the rate of racemisation (Figure 5.1). Typically, at ambient temperature, the 60 MHz ^{1}H NMR spectrum of **10** in the methyl region appears as two sets of doublets indicating that exchange

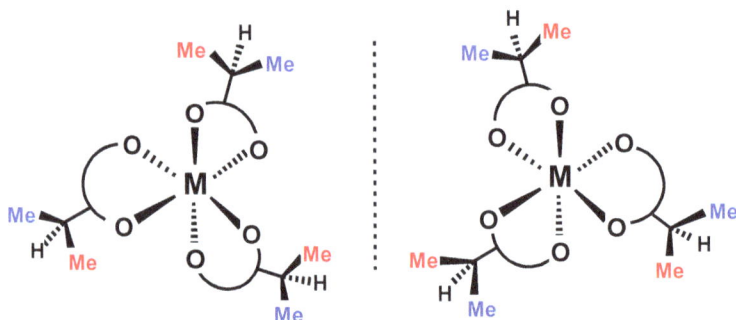

Figure 5.1. Δ and Λ enantiomers of the C_3-symmetric metal-tris(chelate) complex **10** bearing isopropyl substituents possessing diastereotopic methyl groups.

between the enantiomers is still slow on the NMR timescale. However, above 120°C, these peaks coalesce and sharpen, eventually yielding a single well-defined doublet, and the barrier was determined by line-shape analysis to be 22 kcal mol^{-1}.[25]

This approach has since been widely used to probe the stereodynamics of other tris-chelate complexes, as well as many related systems of the type $(L–L)_2MX_2$ and $(L–L)_2MXY$. The various possible mechanisms of loss of stereochemical integrity, either dissociative or via an internal rearrangement such as the Ray-Dutt or Bailar twist, have been comprehensively discussed in a now-classic review and a long series of publications by Serpone and Bickley.[26,27]

Of course, in the pioneering study of Jurado and Springer carried out more than 50 years ago, the available spectrometer operated at 60 MHz for protons. Nevertheless, this was appropriate since on a modern system (operating at 500 MHz or above) the enhanced peak separation in Hz would mean that coalescence would only be observable at a much higher temperature, probably beyond the boiling point of any normal NMR solvent, and possibly above the decomposition point of the complex.

It is therefore important to recognise that for rearrangements with high activation energies one should use the lowest frequency spectrometer consistent with acquiring spectra exhibiting sufficiently well-separated peaks within a reasonable time. In contrast, processes with very low barriers should be studied at the highest possible field so as to bring about the maximum separation (in Hz) between exchanging peaks, otherwise the limiting slow exchange spectrum may not be accessible without recourse to unusual solvent mixtures with very low freezing points. Moreover, when the reactants are thermally sensitive, and raising the temperature until peak coalescence occurs is no longer viable (as in the case discussed in Chapter 3 whereby $(\eta^1$-indenyl$)(\eta^5$-$C_5H_5)Fe(CO)_2$ readily loses two carbonyl ligands to form benzoferrocene), the more recently developed 2D-exchange methods are particularly valuable.[24]

5.4.2. *Metal Cluster Cations*

The remarkable ease of formation and stability of primary carbocations capping a triangle of organo-transition metal fragments, as in

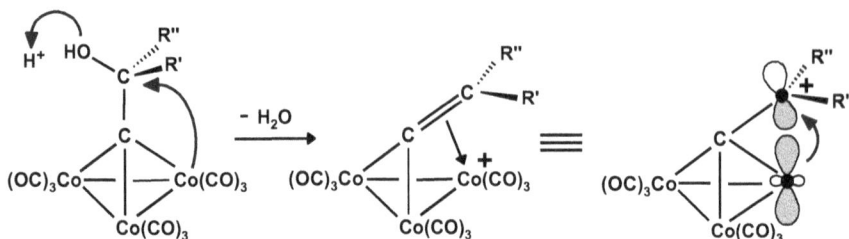

Scheme 5.6.   Formation of a metal-stabilised cation (in **11**, R′ = R″ = H).

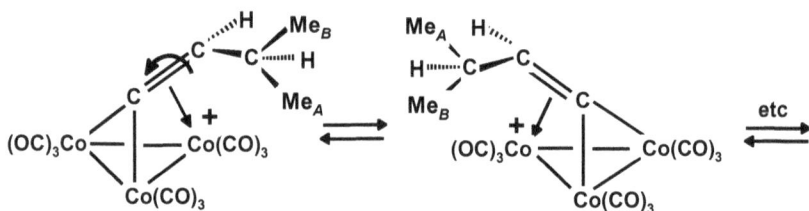

Scheme 5.7.   Antarafacial migration interconverts the diastereotopic methyl groups in **12**.

$[(OC)_9Co_3CCH_2]^+$, **11**, prompted numerous experimental and theoretical studies.[28] Molecular orbital calculations by Schilling and Hoffmann indicated that in the ground state the carbon-carbon linkage of the vinylidene fragment is not aligned with the three-fold axis of the metal triangle, but rather interacts directly with a single cobalt vertex so as to accept electron density from a filled metal d orbital (Scheme 5.6). Moreover, it was suggested that the capping fragment should migrate between the cluster vertices in an antarafacial fashion.[29]

This proposal was brilliantly explored in an ingenious experiment from the Mislow group that verified not only the ground state geometry, but also the details of the fluxional process.[30] The cation **12** contains an isopropyl substituent as a ^{13}C NMR probe for chiral conformations (Scheme 5.7). In the calculated ground state, the cation has C_1 symmetry which would render diastereotopic the methyls of the isopropyl group; gratifyingly, at low temperature, two methyl resonances are indeed observed. Moreover, as the temperature is raised, these two peaks coalesce as would be required for the antarafacial migration pathway via a transition state that possesses a molecular mirror plane; this process equilibrates the two methyl environments, and the measured activation energy was found to be 10.5 kcal mol^{-1}. In this case, the initial broken symmetry verifies the tilted ground state, whereas the recovery of mirror symmetry validates the antarafacial nature of the migration process. As the rearrangement continues, all three tricarbonylcobalt vertices are eventually equilibrated.

5.4.3. *Diphos and Arphos Complexes of Cobalt Clusters*

The potential use of chiral metal clusters as catalysts for asymmetric syntheses inevitably demands that they maintain their stereochemical integrity, and not suffer racemisation under reaction conditions. Tetrahedral carbynyl-tricobalt clusters react with chelating ligands, such as diphosphines, to form complexes of the type $RCCo_3(CO)_7(L–L)$ having C_S symmetry. Use of 1-diphenylarsino-2-diphenylphosphinoethane (arphos), $Ph_2As–CH_2CH_2–PPh_2$, breaks the mirror symmetry and renders the cluster chiral. However, attempts to separate the enantiomers consistently failed, implying that racemisation was occurring.[31]

To monitor this situation, it was necessary to incorporate a diastereotopic probe, and this was accomplished by introducing an isopropyl ester as the substituent on the capping carbyne. At room temperature, the ^1H NMR spectrum of $(arphos)(CO)_7Co_3C–CO_2CHMe_2$, **13**, in the methyl region appears as a doublet (due to splitting by the adjacent hydrogen); evidently, the system racemises at this temperature. However, when cooled to 223 K, this resonance undergoes decoalescence to form two sets of doublets as the diastereotopicity of the isopropyl methyls becomes apparent. Simulation of the variable-temperature spectra yields a relatively modest barrier of 13 kcal mol^{-1}, which accounts for the facile racemisation seen at room temperature.

Somewhat surprisingly, perhaps, the X-ray crystal structure of **13** revealed that the arphos ligand is situated such that the phosphorus and arsenic occupy equatorial rather than axial sites; thus, the mechanism apparently involves fragmentation of the arsenic-cobalt linkage and rotation of the $Co(CO)_2P$ vertex to leave a dangling

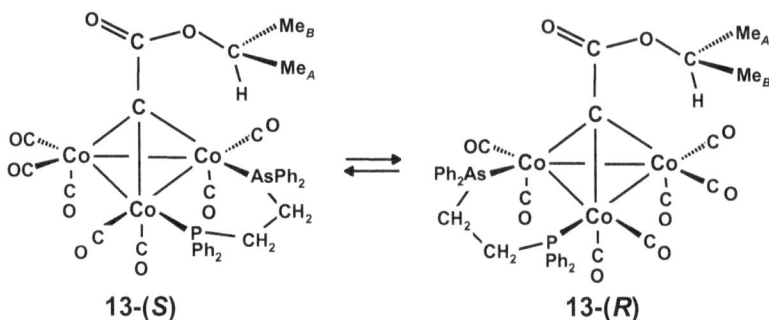

Scheme 5.8.   Migration of the diphenylarseno moiety between cobalt vertices leads to racemisation of the cluster.

diphenylarsino group that can either return to its original position or migrate onto the neighbouring cobalt vertex (Scheme 5.8), thus interconverting *R* and *S* cluster isomers.[31] (The absolute configuration of chiral tetrahedral clusters, such as **13**, can be conveniently designated by assigning an *R* or *S* label to a dummy atom placed inside the tetrahedron, and then assigning priorities in the usual Cahn-Ingold-Prelog manner.[32,33] Thus, in **13** the $(Ph_2As)Co(CO)_2$ vertex has *priority 1*, the $(Ph_2P)Co(CO)_2$ vertex has *priority 2*, the $Co(CO)_3$ vertex has *priority 3*, and the capping carbon atom has the lowest priority.) The rearrangement necessitates concomitant scrambling of the carbonyl ligands, but the barrier to this secondary process was independently measured to be only 10.5 kcal mol^{-1}, and so poses no problem.[31]

   This particular rearrangement has also been investigated using the commercially available enantiopure $1S,2R,5S$-(+)-menthyl substituent to introduce a different source of diastereotopism. Treatment of (+)-menthyl trichloroacetate with dicobalt octacarbonyl, and then with $Ph_2P(CH_2)_2PPh_2$, yields $(diphos)Co_3(CO)_7C$-CO_2(menthyl), **14**, in which the two ^{31}P nuclei are now diastereotopic, and remain so. Evidently, cleavage of the cobalt-phosphorus bond has a higher barrier than for the corresponding cobalt-arsenic linkage. Subsequent replacement of diphos by arphos, as in **15** (Scheme 5.9), introduces a second source of chirality. Once again, at room temperature, interconversion of the resulting diastereomers, via breaking of the Co–As

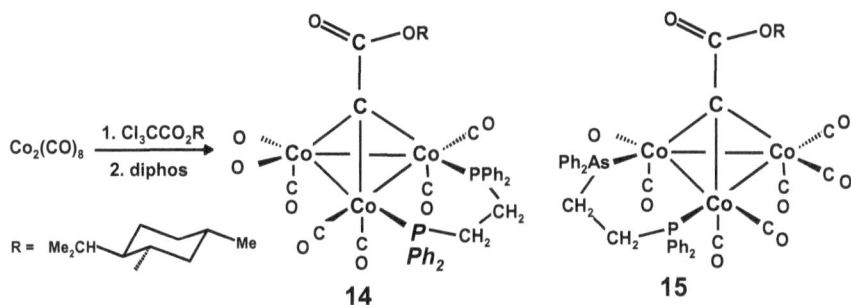

Scheme 5.9. Incorporation of a chiral (+)-menthyl substituent renders the $Co(CO)_2PPh_3$ vertices in **14** diastereotopic; cluster **15** exists as diastereomers.

connection and subsequent migration onto a different cobalt, is rapid and the ^{31}P NMR spectrum appears as a time-averaged singlet. However, at low temperature, peak decoalescence becomes apparent, and a barrier of 13 kcal mol^{-1} was again revealed.[34]

5.4.4. *Mixed Metal Square-pyramidal Clusters*

With the goal of combining the reactivity of metal surfaces with the control achievable in homogeneous systems, especially with regard to asymmetric syntheses, the area of mixed metal clusters has been intensely studied for many years.[35] Clearly, such an undertaking would presuppose that the chirality of the cluster not be lost during the reaction sequence. Typical of such systems are the square-based pyramidal M_3C_2 clusters, in particular those in which the three metal vertices are different and the C_2 unit is derived from an alkyne. As two exemplars, we have selected the clusters $[(C_5H_5)Ni \bullet Ni(C_5H_5) \bullet Fe(CO)_3]$-$[PhC \equiv CCO_2{}^iPr]$, **16**, and $[(C_5H_5)Ni \bullet Co(CO)_3 \bullet Fe(CO)_3]$-$[PhC \equiv CCO_2{}^iPr]$, **17**, in which the isopropyl group acts as a probe for the chiral nature of the cluster.

The unsymmetrical nature of the alkyne moiety in **16** renders the cluster chiral; the (cyclopentadienyl)nickel groups are non-equivalent, as are the diastereotopic methyls of the isopropyl substituent. One can now test for two different processes. One could envisage rotation of the nickel-nickel vector relative to the iron-carbon-carbon triangle

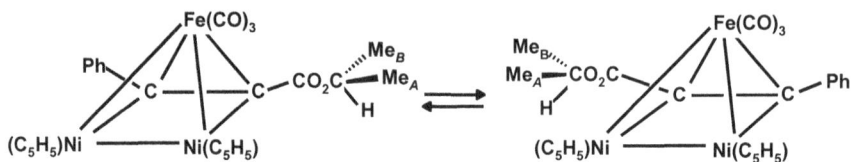

Scheme 5.10. Alkyne rotation racemises the cluster and equilibrates the CpNi vertices in **16**.

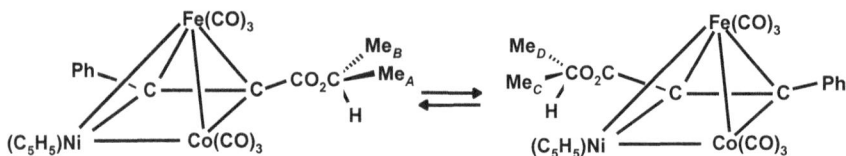

Scheme 5.11. Acetylene rotation only interconverts diastereomers of **17**.

that would interconvert the two nickel environments, but not racemise the molecule; that is, the diastereotopic character of the Me_2CH group would be unaffected. In contrast, a formal rotation of the alkyne moiety relative to the Ni_2Fe triangle not only racemises the molecular cluster, but also interconverts the two $(C_5H_5)Ni$ environments (Scheme 5.10). Were this to be the only rearrangement process, the activation energies for the nickel exchange and coalescence of the methyl peaks must be identical. Gratifyingly, the observed barrier in each case is 15 ± 0.5 kcal mol^{-1}.[36]

One must, however, consider yet another possible occurrence. Dissociation of the cluster into non-complexed alkyne and an unsaturated metal triangle (perhaps weakly stabilised by solvent) would simultaneously equilibrate not only the methyls but also the cyclopentadienyl-nickel environments. It is necessary, therefore, to ensure that the process is indeed intramolecular. This was accomplished by making all three metal vertices different, as in the iron-cobalt-nickel cluster, **17**. As illustrated in Scheme 5.11, acetylene rotation does not interconvert enantiomers but merely diastereomers. Thus, when alkyne rotation is slow on the NMR timescale, each diastereomer will show four peaks in the methyl region of the ^1H spectrum – two magnetically non-equivalent methyls, each doublet split by its adjacent isopropyl hydrogen.

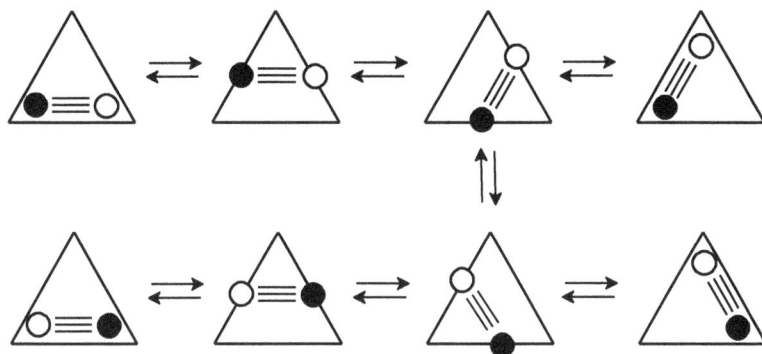

Scheme 5.12. A "modified windscreen-wiper" motion is equivalent to alkyne rotation.

Indeed, at room temperature one sees two major diastereomers, each with its four-peak pattern in the methyl region, but at elevated temperatures there still appears a four-line spectrum indicating that the stereochemical integrity of the chiral cluster is conserved. The experimentally observable barrier of 16.5 ± 0.5 kcal mol^{-1} in the Fe-Co-Ni cluster, **17**, is similar to that observed in the Ni$_2$Fe compound, **16**, indicating that the same mechanistic pathway is involved.[36] Molecular orbital calculations revealed that the rearrangement proceeds via a "modified windscreen-wiper" motion (Scheme 5.12),[37] analogous to that previously proposed for the isolobal $C_5H_5^+$ *nido* cluster.[38,39] Note that simple circumambulation of the alkyne round the periphery of the metal triangle (Scheme 5.13) *does not racemise* the cluster.

5.4.5. *Inversion of Corannulene*

Corannulene, $C_{20}H_{10}$, was first prepared in minuscule yields by Barth and Lawton in a herculean 17-step process;[40] its non-planar 20-carbon framework is a substantial substructure of the fullerene C_{60}. Subsequently, in a synthetic *tour de force*, Larry Scott from Boston College devised a much shorter and more efficient route.[41,42] Moreover, in a very significant advance for the field, Jay Siegel in Zürich modified and developed the synthesis to the point where corannulene is now available in kilogram quantities,[43,44] thus allowing

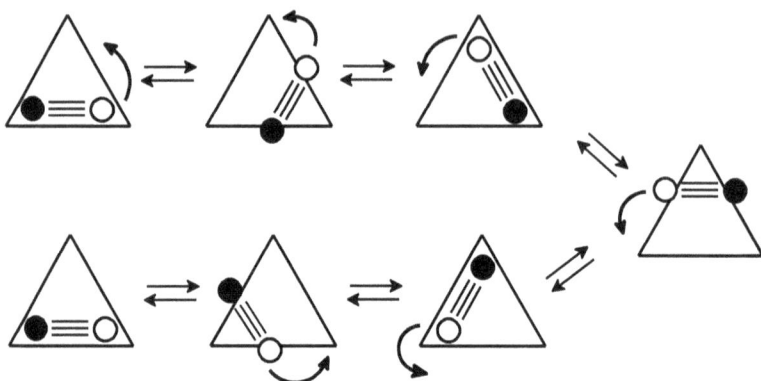

Scheme 5.13. Simple migration of the alkyne around the periphery of the metal triangle does not racemise the cluster.

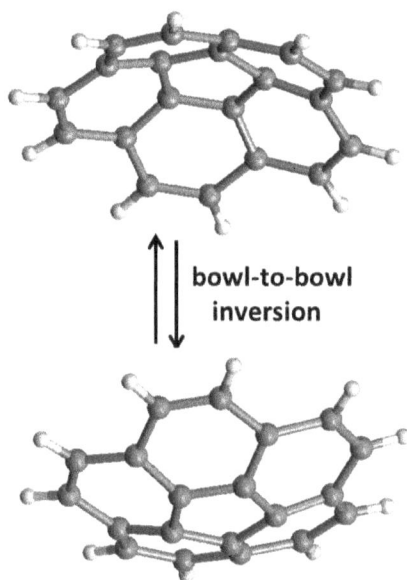

bowl-to-bowl
inversion

Figure 5.2. Corannulene undergoes an umbrella-like process leading to inversion.

the study of its own dynamic behaviour as well as that of numerous derivatives.[45] The barrier to umbrella-like bowl-to-bowl inversion, illustrated in Figure 5.2, involves the interconversion of C_{5v}-symmetric entities via a planar D_{5h} transition state.

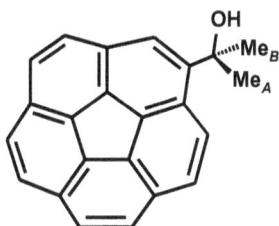

Figure 5.3. The diastereotopic methyl groups in bowl-shaped corannulenyl-dimethylcarbinol, **18**, are equilibrated during the inversion process when the molecule becomes mirror-symmetric.

The activation energy for this process was first experimentally determined from variable-temperature NMR spectroscopic measurements on racemic corannulenyl-dimethylcarbinol, **18** (Figure 5.3). In the slow exchange regime, the diastereotopic methyl groups of the Me_2COH substituent exhibit separate resonances, but they are equilibrated when the C_{20} skeleton adopts a planar geometry; the measured value[46] of 10.2 kcal mol^{-1} may be compared to the DFT-calculated value[47] for corannulene itself of 10.5 kcal mol^{-1}.

5.4.6. *Fluorinated Isopropyl Groups*

As is well known, the chemical shift range of ^{19}F nuclei is very much larger (hundreds of ppm) than that seen in ^1H NMR spectra. Thus, peak separation in ^{19}F spectra can be very large and decoalescence upon cooling the sample can occur at more readily accessible temperatures. For this reason, together with the high receptivity of the ^{19}F nucleus, the incorporation of diastereotopic fluorines into isopropyl groups can be particularly advantageous in systems exhibiting low rearrangement barriers, or when discrimination between structures is rather difficult.

Ishikawa's reagent, *N,N*-diethyl-(1,1,2,3,3,3-hexafluoropropyl) amine, **19**, is now commonly used to convert alcohols into alkyl fluorides. However, since it possesses diastereotopic fluorines, these are also useful to probe conformational preferences (Scheme 5.14) by monitoring the changing value of $^3J_{HF}$ in a range of different solvents,[48] analogous to the widely used Karplus curve that correlates the magnitude of $^3J_{HH}$ with the dihedral angle between coupled nuclei.[49]

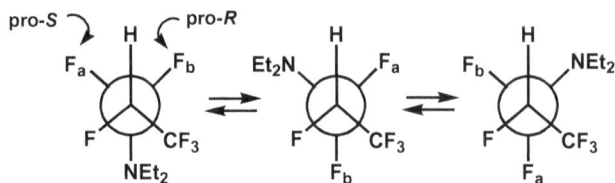

Scheme 5.14. Structure of Ishikawa's reagent, **19**, and Newman projections of rotamers yielding different values of $^3J_{HF}$.

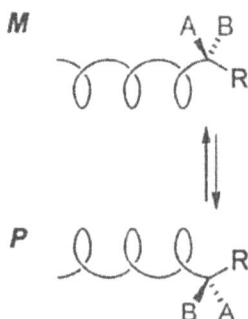

Figure 5.4. Schematic representation of helical foldamers of opposite screw-sense bearing diastereotopic labels, where **A** and **B** are CH_2F fragments.

A particularly elegant example of the use of diastereotopic ^{19}F nuclei comes from Jonathan Clayden's group at the University of Manchester who utilised this phenomenon to probe the screw-sense preference of helical foldamers. In this particular case, helical oligomers of the quaternary achiral amino acid Aib (2-aminoisobutyric acid) undergo screw-sense inversion with half-lives on the order of 10^{-4} secs, a timescale appropriate for an NMR investigation. Attachment of the amino acid ester $H_2N–C(CH_2F)_2–CO_2CH_3$, at the C-terminus of the helical oligamers (shown schematically as Figure 5.4) renders the fluoromethyl groups diastereotopic, and allows the measurement of helical excess and the rate of their interconversion.[50] (The designations **M** (minus) and **P** (plus) indicate the rotational sense (left- or right-handed) of the helix.)

In very recent work, a number of compounds of the general struc-ture aryl-$CF(CF_3)CH_2F$, possessing an enantioenriched pentafluori-nated isopropyl group, have been prepared via iodine(I)/(III) catalysis, and has been used to expand the portfolio of fluorinated systems available for drug discovery.[51]

5.5. Metal Complexes of Hexaethylbenzene and Related Systems

Hexaethylbenzene (HEB) is the smallest homo-substituted hexaalky-lbenzene in which the effects of steric crowding are manifested.[52] It adopts a structure (Figure 5.5) whereby alternate alkyl groups lie above and below the plane of the arene, giving rise to D_{3d} symmetry that renders all six ethyl groups NMR equivalent.[53]

Attachment of a π-bonded metal tricarbonyl fragment, as in $(\eta^6\text{-}C_6Et_6)Cr(CO)_3$, **20**, breaks the symmetry between the faces of the arene ring such that the ethyl groups are now separated into *prox-imal* and *distal* environments, and lowers the point group to C_{3v}, as was first shown in a solid-state ^{13}C NMR study.[54] In solution, how-ever, the ^{1}H and ^{13}C spectra of **20** at ambient temperature revealed that all the ethyl groups were equivalent, indicating a "time-aver-aged" C_{6v} structure where both ethyl and tripodal rotation were fast on the NMR timescale (Scheme 5.15).[53] Analogous dynamic behav-iour was also found in $(\eta^6\text{-}C_6Et_6)W(CO)_3$.[55]

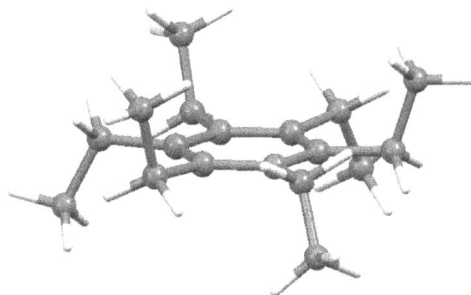

Figure 5.5.   The D_{3d} structure of hexaethylbenzene.

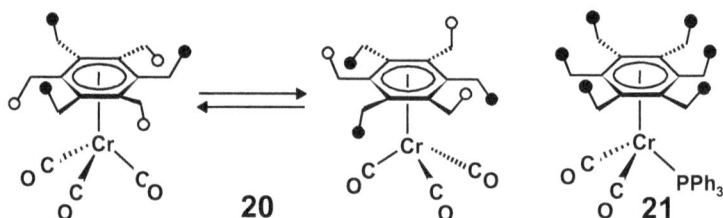

Scheme 5.15. Interconversion of *distal* and *proximal* ethyls in $(HEB)Cr(CO)_3$, **20**, combined with rapid tripodal rotation, generates effective C_{6v} symmetry, whereas the bulky PPh_3 ligand in **21** brings about a conformational change such that all six ethyls are *distal*.

As the temperature is lowered, ethyl rotation slows to the point such that the alkyl substitutents are eventually "frozen out" on the NMR timescale into *proximal* and *distal* sets relative to the π-bonded $Cr(CO)_3$ fragment. *However, it is not possible to tell whether or not tripodal rotation has slowed since the system retains its C_{3v} symmetry in either case.* An attempt to enhance the hindrance to rotation by incorporation of a bulky triphenylphosphine group into the tripod, as in $(\eta^6\text{-}C_6Et_6)Cr(CO)_2PPh_3$, **21**, led instead to a different conformer whereby the six ethyls are always *distal*, and the NMR spectrum is unchanged upon lowering the temperature.[53]

Evidently, a more subtle change in molecular symmetry was required, and this was first reported by McGlinchey *et al.* Replacement of one of the carbon monoxide ligands by the structurally similar, cylindrical thiocarbonyl moiety lowers the symmetry to C_S in $(\eta^6\text{-}C_6Et_6)Cr(CO)_2CS$, **22**, and splits the ethyl groups into a 2:1:2:1 pattern when tripodal rotation is slow.[55,56] Even more striking, introduction of a third different ligand, a nitrosyl group, as in $[(\eta^6\text{-}C_6Et_6)Cr(CO)(CS)NO]^+[BF_4]^-$, **23**, completely breaks the degeneracy of the six ethyl substituents and results in an 18-peak ^{13}C NMR spectrum (shown in Figure 5.6) since, in the point group C_1, each of the six methyl, methylene and ring carbon environments is now different (Scheme 5.16).[57] One could envisage the situation whereby the two dynamic processes, ethyl rotation and tripodal rotation, proceed in a correlated fashion analogous to a conical gear. However, simulation of the spectra revealed that these two barriers were clearly different,

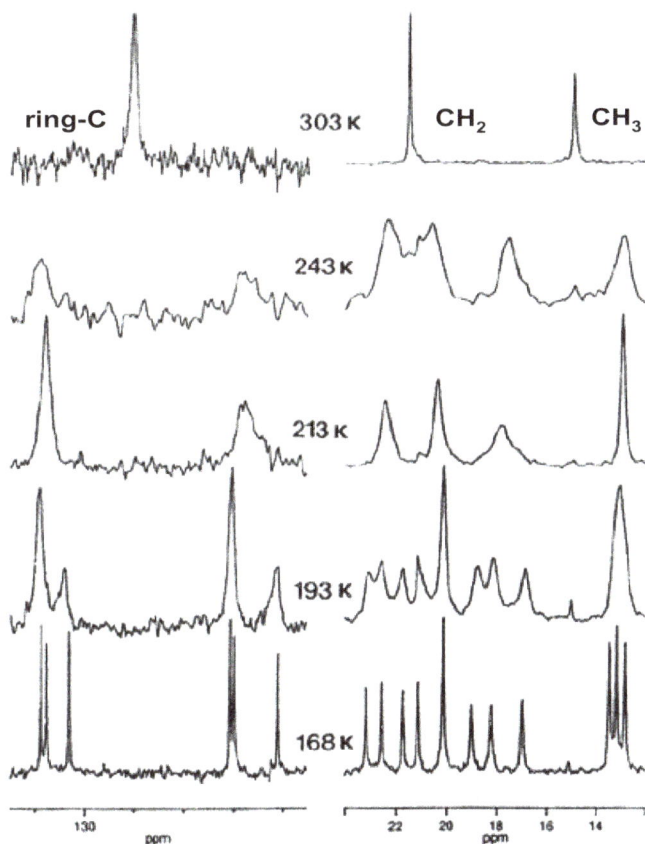

Figure 5.6. Variable-temperature ^{13}C NMR spectra of $[(\eta^6\text{-}C_6Et_6)Cr(CO)(CS)$ $NO]^+$, **23**.

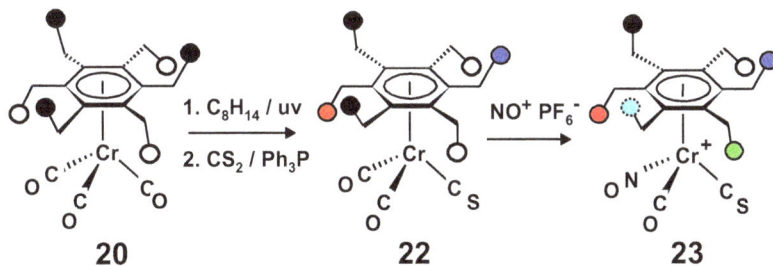

Scheme 5.16. Sequential replacement of a carbonyl ligand by thiocarbonyl and nitrosyl yields complexes **20**, **22** and **23** with C_{3v}, C_S and C_1 symmetry, successively.

11.5 and 8.5 kcal mol^{-1}, for ethyl and tripodal rotation, respectively.[57] It is therefore evident that these processes occur independently, not in a correlated fashion.

Of course, it is important that substitution to lower the molecular symmetry does not incur any significant electronic or steric effect that would unduly perturb the system. To this end, in the HEB-chromium system, where carbonyls were sequentially replaced by thiocarbonyl and nitrosyl ligands, compounds **20**, **22** and **23** were each characterised by X-ray crystallography to ensure consistency of the molecular geometry.[53,55,57]

Analogous dynamic behaviour, in each case yielding a tripodal barrier of ~9.4 kcal mol^{-1}, has also been observed in other C_s-symmetric systems such as $(HEB)Cr(CO)_2-N=N-Cr(CO)_2(HEB)$, [1,3,5-triethyl-2,4,6-tris(trimethylsilylmethyl)benzene]$Mo(CO)_2CS$ and $(HEB)Mn(CO)_2Br$.[52] The barrier to tripodal rotation in molecules of the type $(arene)M(CO)_3$ is of significance since it has been proposed that the orientation of the tripod directs the site of attack at the arene ring by an incoming nucleophile. Consequently, approaches towards controlling the orientation of the tripod, either sterically or electronically, have been widely investigated.[58]

Although the dynamic behaviour of HEB itself has not been directly elucidated experimentally, the possibility of invoking a correlated process whereby all six ethyl groups rotate simultaneously was shown to be non-viable on both energetic and entropic grounds. The most reasonable proposal for HEB itself involves the six-step sequence $(a) \rightarrow (b) \rightarrow (c) \rightarrow (d) \rightarrow (c) \rightarrow (b) \rightarrow (a)$ whereby the three *distal* and three *proximal* substituents exchange positions in the most expeditious manner (Scheme 5.17).[59]

The question of slowed rotation of a π-bonded chromium tricarbonyl fragment has also been investigated by breaking the symmetry of the hexa-substituted arene rather than by rendering chiral the tripodal moiety. Replacement of one ethyl group by an acetyl substituent, as in pentaethylacetophenone-tricarbonylchromium, **24**, leaves the steric environment of the $Cr(CO)_3$ group essentially unchanged, but slowed tripodal rotation splits the three-fold degeneracy such that the ^{13}CO NMR resonances now exhibit a 2:1 pattern. As with the HEB

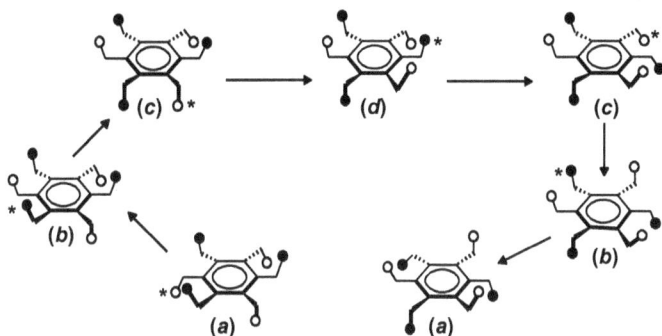

Scheme 5.17. The lowest energy six-step pathway that interconverts ethyl groups on opposite faces of HEB. (In each case, the substituent due to rotate next is marked with an asterisk.)

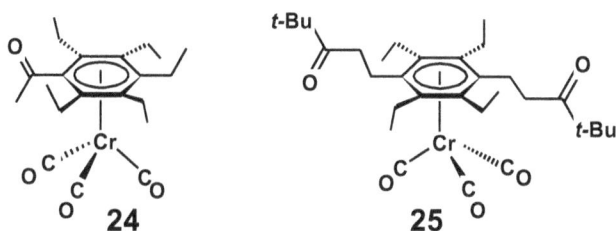

24

25

Figure 5.7. Complexes in which the six-fold symmetry of the arene ring has been broken.

complexes **20**, **22** and **23**, the tripodal rotational barrier in **24** was again 9.5 kcal mol^{-1}.[60]

Another approach to measuring the barriers to alkyl and tripodal rotations in $(\eta^6\text{-}C_6R_6)Cr(CO)_3$ systems was reported in a particularly elegant parallel investigation by Kilway and Siegel.[61] In this case, the arene chosen was [1,4-bis(4,4-dimethyl-3-oxopentyl)-2,3,5,6-tetra-ethylbenzene] tricarbonylchromium, **25** (Figure 5.7). In this molecule, with its alternating up-down pattern of substituents, the two ketones are situated on opposite faces of the central ring, one *distal* the other *proximal*. When alkyl and tripodal rotations are slowed on the NMR timescale, the molecule has only mirror symmetry (C_S), thus splitting the $Cr(^{13}CO)_3$ resonance into a 2:1 pattern. As the temperature is gradually raised, the chromium carbonyl resonances coalesce into a singlet, but the two ketonic environments remain non-equivalent until the onset of alkyl rotations, at which point the molecule has effective C_{2v} symmetry. The barriers to ethyl and tripodal rotation were found to be 11.8 and 9.5 kcal mol^{-1}, respectively, in excellent agreement with the earlier work.[57]

In related purely organic molecules, the favoured conformation of hexa-*n*-propylbenzene parallels the fully alternating "up-down" (D_{3d}) structure of HEB, whereas in hexa-*iso*propylbenzene, **26**, the alkyl groups form a tightly interlocked cyclic tongue-and-groove arrangement (Figure 5.8) such that they are all pointing in the same direction, giving rise to C_{6h} symmetry. The exceptional rigidity of this system is reflected in the high barrier to homomerisation of at least 22 kcal mol^{-1}, as shown by specific deuteration of two methyl groups, in conjunction with Empirical Force Field calculations.[62] Similar behaviour was also seen in hexakis(dichloromethyl)benzene.[63]

Considering now a related organometallic system, Gloaguen and Astruc[64] from the University of Bordeaux prepared the very crowded cobalt sandwich cation $[(C_5{}^iPr_5)Co(C_5H_5)]^+$, **27**. Once again, the isopropyl substituents are arranged such that they all point in the same direction, thus giving rise to C_5-symmetric enantiomers (Figure 5.9).

At room temperature the ^{1}H and ^{13}C NMR spectra of **27** reveal the presence of two methyl environments, *exo* and *endo*, that only coalesce at 100°C indicating a substantial barrier to racemisation. The methyl protons are rendered diastereotopic by the helicity of the

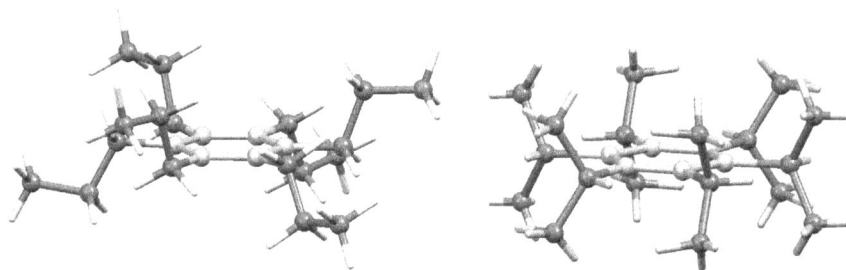

Figure 5.8. Structures of (left) hexa-*n*-propylbenzene and (right) hexa*iso*propylbenzene, **26**.

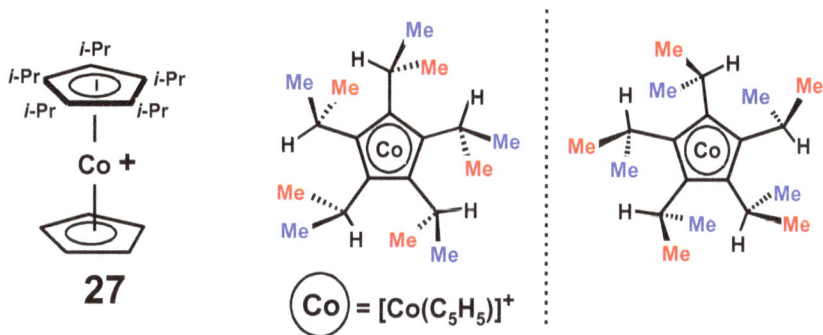

Figure 5.9.   Enantiomers of $[(C_5{}^iPr_5)Co(C_5H_5)]^+$, **27**, arising from restricted rotation of the isopropyl substituents.

penta*iso*propylcyclopentadienyl ring that senses the clockwise and counter-clockwise orientations of the gear-meshed substituents. However, the smaller ring size of the cobalt sandwich complex **27** compared to $C_6{}^iPr_6$ widens the angles between the isopropyl groups allowing more space than in **26**, and the rotational barrier is reduced to 17 kcal mol^{-1}.[64] Once again, it is incorporation of a π-complexed moiety, $Co(C_5H_5)$, that breaks the C_{5h} symmetry of the $C_5{}^iPr_5$ ring and allows the detection and measurement of the barrier to alkyl rotation.

5.6. Rotations of Peripheral Ring Substituents in C_nAr_n and Related Metal Complexes

5.6.1. *Hexa-arylated Benzenes*

Perphenylated ring systems, C_nPh_n, where n ranges from 3 through 7, adopt propeller conformations whereby the twist angles made by the peripheral rings with respect to the central ring reflect the compromise between electronically favoured conjugative coplanarity and the ugly reality of steric hindrance. As the ring size increases, the angle ω subtended by adjacent moieties at the centre of the internal ring decreases: C_3Ph_3 120°, C_4Ph_4 90°, C_5Ph_5 72°, C_6Ph_6 60°, C_7Ph_7 51.4°. Although increasing the ring size lengthens the radial distance of the external groups from the ring centre, this is more than compensated for by the diminishing value of ω. As a result, the peripheral rings find themselves in an increasingly crowded locale such that the twist angles for $[C_5Ph_5]^-$, C_6Ph_6 and $[C_7Ph_7]^+$ are approximately 50°, 70° and 80°, respectively.[65,66] Indeed, the heptaphenyltropylium cation, $[C_7Ph_7]^+$, is not planar and adopts a pronounced boat conformation.[67]

These systems have been heavily investigated as molecular analogues of propellers or gears.[65,68] They adopt D_n conformations whose peripheral phenyl rotations cannot be monitored unless the symmetry is lowered, such as by incorporation of *ortho* or *meta* substituents into these rings, or by π-complexation to an organometallic moiety.

The dynamic behaviour of hexaarylbenzene systems has been elucidated by Devens Gust at Arizona State University. It was shown that, on the NMR timescale, these molecules exist in a conformation in which they are perpendicular to the plane of the central ring. However, when bearing suitable substituents, they display restricted rotation with barriers ranging from ~33 kcal mol^{-1} for methoxy groups in *ortho* positions to ~17 kcal mol^{-1} for methyls in *meta* positions.[69] Inversion of molecular propellers requires rotation of all substituted rings through approximately π radians, thus leading to enantiomerisation, as in Scheme 5.18. However, this occurs by sequential rotation of individual peripheral rings since correlated rotation of all six substituents is energetically and entropically disfavoured.

While the most obvious steric interactions are with the substituents neighbouring the one undergoing rotation, these effects are

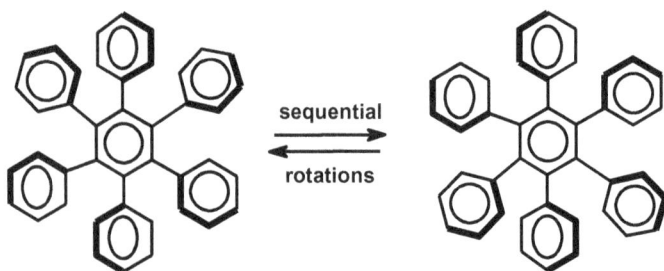

Scheme 5.18. Racemisation of hexaarylbenzenes proceeds via a series of single rotations.

spread throughout the molecule. However, calculations at the DFT level, whereby a single ring was rotated gradually, revealed that the *meta* substituent responded most strongly as the process developed. This phenomenon is particularly noticeable in hexa(β-naphthyl)benzene, **28**, that was prepared by the Diels-Alder reaction of di(β-naphthyl)acetylene with tetra(β-naphthyl)cyclopentadienone, wherein the external benzo ring is effectively a *meta*-substituent with an augmented wingspan, and the naphthyl rotational barrier was found to be approximately 17 kcal mol^{-1}.[70]

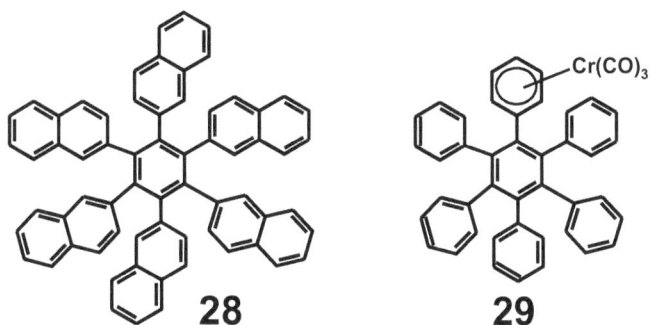

28　　　　　**29**

5.6.2. *Organometallic Derivatives of C_nAr_n Systems*

The symmetry of hexaphenylbenzene has also been broken by the incorporation of a π-complexed organometallic moiety in (C_6Ph_6) $Cr(CO)_3$, **29**. In that case, rotation of the chromium-complexed ring was measured as 12.2 kcal mol^{-1},[71] considerably lower than the values found in the *ortho-* and *meta*-substituted hexaarylbenzene systems studied by Gust. Numerous examples of arenes bearing multiple organometallic substituents have been reported. Typically, 1,3,5-triphenylbenzene bearing one, two or three tricarbonylchromium groups, **30**, on the peripheral rings have been prepared by direct complexation with the arene and characterised spectroscopically or by X-ray crystallography;[72] similarly, 1,3,5-substituted benzenes bearing three ferrocenyl or (cyclopentadienyl)manganesetricarbonyl groups are readily available by trimerisation of the precursor methyl ketones with tetrachlorosilane in ethanol, as exemplified by the mixed Fe$_2$Mn complex **31** (Figure 5.10).[73,74]

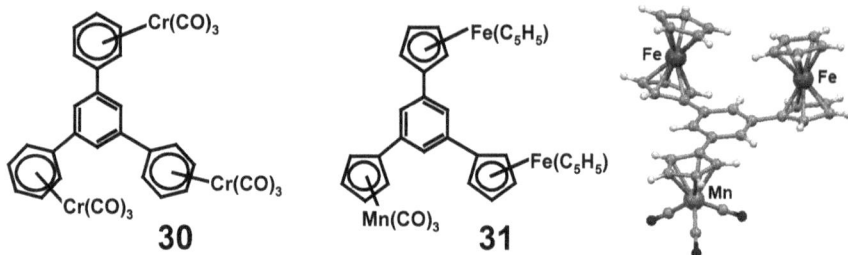

Figure 5.10. (1,3,5-triphenylbenzene)[$Cr(CO)_3$]$_3$, **30**, and the X-ray crystal structure of 1,3-diferrocenyl-5-cymantrenyl-benzene, **31**.

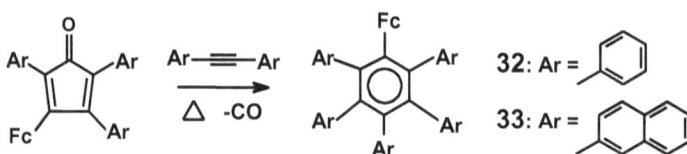

Scheme 5.19.   Syntheses of ferrocenyl-pentaphenyl-, **32**, and -pentanaphthyl-benzene, **33**.

The hexaarylbenzene complexes discussed above adopt propeller conformations in the solid state with dihedral angles of 65–70° relative to the central ring. However, replacement of a peripheral ring in hexaphenyl- or hexa(β-naphthyl)benzene by a ferrocenyl unit has been accomplished via Diels-Alder reaction of the appropriate ferrocenylcyclopentadienone and diarylacetylene (Scheme 5.19). In those cases, the structure is not a propeller but rather an incremental progression of twist angles such that the phenyls adjacent to the ferrocenyl moiety in **32** make dihedral angles of +64° and –60° with the central ring; in ferrocenylpenta(β-naphthyl)benzene, **33**, the corresponding angles are ±70°.[75,76]

Finally, we note the successful synthesis of the long-sought hexaferrocenylbenzene, **34**, by the always brilliantly creative Peter Vollhardt at Berkeley. This molecule was finally obtained by the six-fold Negishi-type palladium-catalysed ferrocenylation of hexabromo- or hexaiodobenzene by using diferrocenylzinc (Scheme 5.20). In the solid state, the ferrocenyls project alternately above and below the central ring with dihedral angles of +31° and –82°. Likewise, in

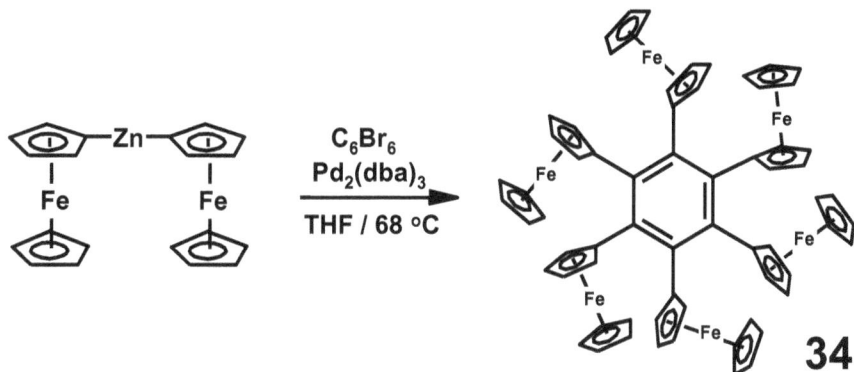

Scheme 5.20.   Synthesis of hexaferrocenylbenzene, **34**.

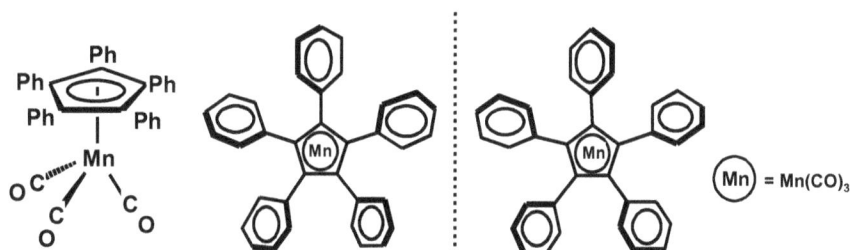

Figure 5.11.   Enantiomers of $(\eta^5\text{-}C_5Ph_5)Mn(CO)_3$ arising from the clockwise and counter-clockwise orientations of the peripheral phenyl substituents.

solution at $-60°C$, decoalescence of the ^1H NMR signals in the C_5H_4 ring are also in accord with the alternating "up-down" structure, analogous to that already seen in HEB; the rotational barrier was found to be 10.5 ± 0.6 kcal mol^{-1}.[77]

In the case of a non-chiral tripod, typified by $(\eta^5\text{-}C_5Ph_5)$ $Mn(CO)_3$,[78] rapid rotation of phenyls and of the tripod generates time-averaged C_{5v} symmetry, but separation into clockwise and counter-clockwise orientations of the peripheral phenyls, as in Figure 5.11, gives rise to enantiomers of C_5 symmetry.[79]

However, when phenyl rotation (or even oscillation) becomes slow on the NMR timescale, the presence of a chiral tripod, as in $(\eta^5\text{-}C_5Ph_5)$ $Fe(CO)(CHO)PMe_2Ph$, **35**, provides another stereocentre and generates C_1-symmetric diastereomers (Figure 5.12). This situation renders

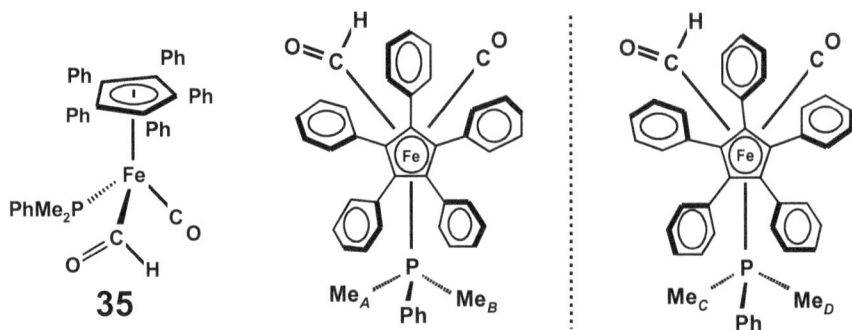

Figure 5.12. Diastereomers of **35**, arising from opposite orientations of the five-bladed propeller.

the phosphorus nuclei diastereotopic, thus giving rise to two ^{31}P NMR resonances whose decoalescence behaviour allows the evaluation of the barrier to interconversion of the C_5Ph_5 helices as 11.7 ± 0.3 kcal mol^{-1}. Moreover, within each molecule, the two methyl groups on the phosphine ligand are also diastereotopic, giving rise to a pair of signals; hence, when interconversion of the C_5Ph_5 five-bladed propellers becomes slow on the NMR timescale, we see *two sets* of diastereotopic methyl pairs.[80]

On a 500 MHz spectrometer at 173 K, slowed tripodal rotation splits the five-fold degeneracy of the C_5Ph_5 moiety in **35** and yields a barrier of 8.7 ± 0.3 kcal mol^{-1}. It is evident that the marked difference between the activation energies of these two processes shows that they are *not* correlated; indeed, the tripod continues to spin rapidly even when rotation of the peripheral phenyls has become slow on the NMR timescale. The net result is that sequentially enhanced symmetry breaking allows complete determination of the stereodynamic behaviour of the system.

5.6.3. *^{19}F NMR Spectroscopy as a Probe for Hindered Aryl Rotations*

We have already encountered barriers to phenyl rotations in systems of the type $(C_nPh_n)ML_x$, where the π-bonded organometallic substituent breaks the D_n symmetry of the free ligand by differentiating between the faces of the central ring. However, there are other cases where the aryl group is sufficiently sterically crowded to exhibit hindered rotation without the need to incorporate an additional bulky group.

For example, the reaction of pentafluorophenyllithium with tetraphenylcyclopentadienone yields (among other isomers) 5-pentafluorophenyl-1,2,3,4-tetraphenylcyclopentadien-5-ol, **36**. Its X-ray crystal structure reveals the crowded environment of the pentafluorophenyl ring such that it is oriented orthogonal to the plane of the central ring, thus splitting the degeneracy of the *ortho*, and also the *meta*, fluorines. In this system, slowed rotation of the pentafluorophenyl ring is evident even at room temperature, as indicated by the appearance of five ^{19}F NMR resonances — a phenomenon greatly enhanced by the much larger chemical shift dispersion exhibited by fluorine nuclei in aromatic rings compared to that shown by protons in the analogous situation. The data shown in Figure 5.13 were acquired at 282 MHz on a 300 MHz spectrometer, and the fluorine resonances are spread over more than 20 ppm (~6500 Hz); on a 500 MHz NMR spectrometer, where fluorine spectra are run at 471 MHz, separation would exceed 10,000 Hz. The barrier to rotation of the fluorinated ring in **36** was evaluated as 20 kcal mol^{-1}.[81]

A more challenging problem arose when it was predicted from DFT calculations that the barrier to phenyl rotation in

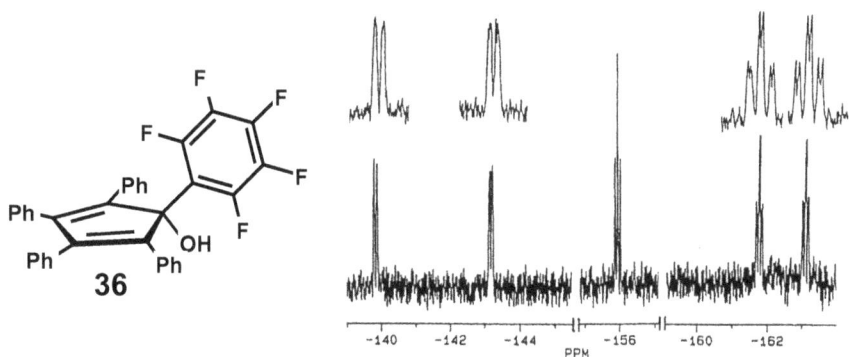

Figure 5.13. 282 MHz ^{19}F NMR spectrum of 5-pentafluorophenyl-1,2,3,4-tetraphenylcyclopentadien-5-ol, **36**, showing five fluorine environments at 30°C.

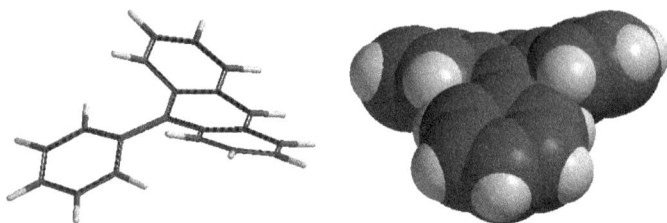

Figure 5.14.   The DFT-calculated non-planar transition state structure of 9-phenyl anthracene, **37**, as the phenyl attempts to rotate past the plane of the anthracene unit.

9-phenylanthracene, **37**, was also 20 kcal mol^{-1}.[82,83] The steric hindrance arises because the two ring systems cannot become coplanar without bringing the phenyl *ortho* hydrogens too close to H(1) and H(8) of the anthracene skeleton. The DFT-calculated transition state revealed a non-planar, stepped arrangement, in which the anthracenyl fragment is markedly curved, as shown in Figure 5.14.[84]

The intrinsic C_{2v} or D_{2h} symmetry of 9-phenylanthracene or 9,10-diphenylanthracene, respectively, renders equivalent the pairs of *ortho* and *meta* CH positions in the phenyl rings and so does not allow the rotational process to be monitored by variable-temperature NMR spectroscopy. Hence, it is necessary to break the symmetry, but to do so in such a fashion so as not to perturb the molecular geometry significantly. This was accomplished in two ways. Initially, 9,10-bis(3-fluorophenyl)anthracene, **38**, was synthesised and its variable-temperature NMR spectra were acquired. The phenyl ring carbons exhibited multiple ^{13}C resonances indicating clearly the presence of *syn* (C_{2v}) and *anti* (C_{2h}) rotamers; more significantly, however, one sees two fluorine environments in the ^{19}F NMR spectrum in solution, and the barrier to their interconversion was found to be 21 kcal mol^{-1}. Subsequently, the X-ray crystal structure revealed the presence of equal populations of *syn* and *anti* isomers; the structure of *anti*-**37** appears as Figure 5.15, and reveals that the fluorophenyls adopt an angle of 85° with the plane of the anthracene thus deviating only slightly from the perpendicular orientation.[84]

The next step was to investigate the rotational barrier of an unsubstituted phenyl. This requires that the symmetry of the *environment* of the phenyl has to be broken rather than the symmetry of the

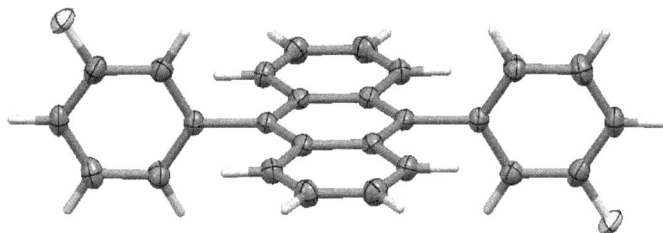

Figure 5.15.   X-ray crystal structure of *anti*-9,10-bis(3-fluorophenyl)anthracene, **38**.

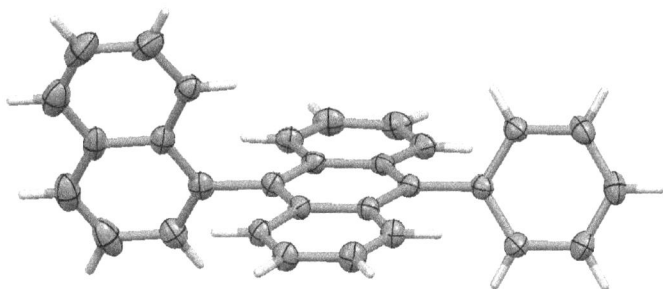

Figure 5.16.   X-ray crystal structure of 9-(1-naphthyl)-10-phenylanthracene, **39**.

phenyl itself. To this end, 9-(1-naphthyl)-10-phenylanthracene, **39**, was prepared and characterised by X-ray crystallography; the molecule adopted an almost mirror-symmetric structure such that the phenyl and naphthyl rings made dihedral angles of 80° and 88°, respectively, with the plane of the anthracene (Figure 5.16). Gratifyingly, the 600 MHz ^{1}H NMR spectrum of **39** exhibited five resonances for the phenyl ring, even at room temperature, and peak coalescence behaviour yielded a phenyl rotational barrier of 21 kcal mol^{-1}.[84] Once again, judicious symmetry breaking allowed the detection and measurement of the barrier to a hidden molecular rearrangement process.

5.7. Rotations and Migrations in Indenyl and Related Systems

In Chapter 3, in the context of the Woodward-Hoffmann rules, we have already discussed the migration of σ-bonded organosilyl groups, as well as $(C_5H_5)Fe(CO)_2$ moieties, across the surface of cyclopenta-dienyl and indenyl rings. We here focus on the dynamic behaviour of metal-alkene units π-bonded to indenyl and related molecular frameworks.

5.7.1. *Indenyl and Ethylene Rotations in (Indenyl)*
Bis(ethylene)rhodium(I)

The use of NMR spectroscopy to measure rotational barriers in met-al-alkene complexes dates back 50 years to the classic report by Cramer[85] in which he noted that the alkene protons in $(\eta^5\text{-}C_5H_5)$ $Rh(C_2H_4)_2$, **40**, which has effective C_{2v} symmetry, could be distin-guished by their relative orientations with respect to the cyclopenta-dienyl ring. Thus the "outside" (δ 2.86) and "inside" (δ 1.03) protons could, in principle, be interconverted by a formal rotation about an axis joining the rhodium to the centre of the carbon-carbon double bond. Variable-temperature NMR studies elucidated this bar-rier to alkene rotation as 15 kcal mol^{-1}.[86]

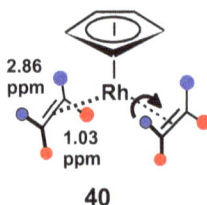

40

In contrast, in the analogous indenyl complex $(\eta^5\text{-}C_9H_7)$ $Rh(C_2H_4)_2$, **41**, the corresponding barrier was reported to be consid-erably lower, 10.5 kcal mol^{-1}. It was suspected that the ability of the indenyl ligand to slide from an η^5- to an η^3-bonding mode, thus gen-erating aromatic character in the six-membered ring, could be rele-vant. The 60 MHz ^{1}H NMR spectrum of **41** shows, in the ethylene region, a single resonance (doublet split by ^{103}Rh) at 50°C, which splits into two multiplets at –90°C. This was rationalised in terms of slowed ethylene rotation on the NMR timescale with a barrier of 10.3 kcal mol^{-1}.[87,88]

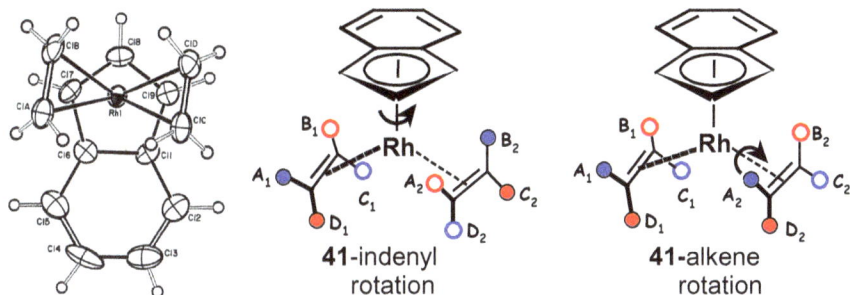

Figure 5.17. Indenyl and alkene rotations interconvert different pairs of alkene protons in (indenyl)bis(ethylene)rhodium(I), **41**.

However, one must exercise some caution since in **41** the symmetry is lowered from effective C_{2v} to C_S, and there are *four*, not *two*, proton environments in each C_2H_4 ligand. Complete equilibration of these protons requires not only rotation about the alkene-rhodium axis, but also rotation of the $Rh(C_2H_4)_2$ fragment about the metal-indenyl axis. Figure 5.17 shows the X-ray crystal structure of **41**, and also the two rotation mechanisms that interconnect different proton environments.

Specifically, rotation about the metal-indenyl vector interconverts A_1 and B_2, B_1 and A_2, C_1 and D_2, and D_1 and C_2. In contrast, alkene rotation interconverts A_1 and C_1, B_1 and D_1, A_2 and C_2, and B_2 and D_2. *Thus, the A_1 proton can gain access only to the B_2, C_1 and D_2 positions, and it is only the presence of a molecular mirror plane that renders A_1 equivalent to A_2 (and thence to B_1, C_2 and D_1).* In the absence of a molecular mirror plane the eight protons will fall into two non-interconvertible sets of four. In this case, each of these two sets of four is made up of a *trans*-related pair of protons from different ethylenes; we do not simply have diastereotopic ethylenes each providing a set of four interconvertible proton environments.[89]

To monitor these two rotational processes independently, it is necessary to break the molecular mirror plane, as in (1-methylindenyl)bis(ethylene)rhodium(I), **42**, which, at 165 K on a 500 MHz spectrometer, indeed exhibits four ^1H NMR resonances in the ethylene "outside" region (3.2–2.2 ppm) and, likewise, four more in the "inside" region (1.5–0.8 ppm); each of these absorptions is, of course, doublet split by ^{103}Rh with J_{Rh-H} ~ 2.5 Hz. Experimentally, the initial coalescence from 165 K to 200 K interchanges proton environments only *within* the "outside" and *within* the "inside" regions. This is ascribed to rotation of the Rh(C$_2$H$_4$)$_2$ fragment about the metal-indenyl axis, and yields a barrier of 8.5 ± 0.4 kcal mol^{-1}. Subsequently, as one approaches room temperature, coalescence between the "inner" and "outer" sets leads ultimately to two resonances, and the barrier was found to be 10.5 ± 0.5 kcal mol^{-1}. This latter process must be assigned to ethylene rotation which clearly has a higher barrier than indenyl rotation, showing that these two rearrangements are not correlated.[89] Overall, one can conclude that the incorporation of a methyl group into the five-membered ring in **42** not only breaks the mirror symmetry, thus allowing independent monitoring of indenyl and alkene rotations, but also maintains the separation of the eight ethylene protons into two sets of four.

This has an interesting consequence for the ^{13}C NMR spectra of **42** which exhibit four alkene peaks at 165 K, two peaks at 220 K and a single resonance at 280 K when the combination of the two rotation processes allows each ethylene carbon access to all three other sites (Scheme 5.21). In particular, we note that the two peaks observed at 200 K do not arise from individual ethylenes, but rather each peak comprises a pair of carbons *from different ethylenes* but related by the two-fold rotation about the indenyl-rhodium axis.[89]

An analogous study on bis(trimethylphosphine)(1-methylindenyl) rhodium(I), **43**, in which the ^{31}P nuclei are diastereotopic, allowed

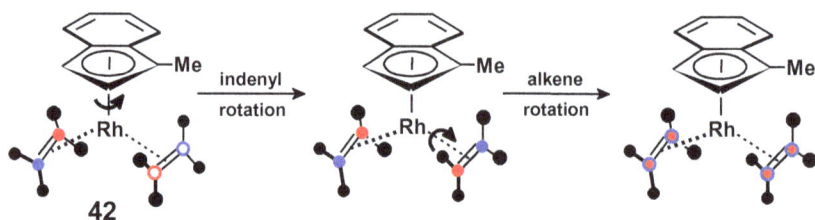

Scheme 5.21.   Indenyl and alkene rotations interconvert different pairs of alkene carbons in (1-methylindenyl)bis(ethylene)rhodium(I), **42**.

direct evaluation of the barrier to rotation of the $Rh(PMe_3)_2$ fragment about the indenyl-rhodium axis as 11.2 kcal mol^{-1}.[90]

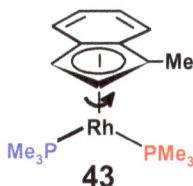

43

5.7.2. *Norbornadiene-rhodium Complexes of Pentamethylcorannulene*

One should note an elegant extension of these concepts by the Siegel/Baldridge group whereby alkene rotation and migration of a metal-diene across the corannulene surface has been investigated.[91] As noted above in Section 5.4.5, corannulene undergoes rapid bowl-to-bowl inversion; however, when substituted at the 1,3,5,7,9 positions, the two forms are enantiomeric. Coordination of a cationic (η^4-norbornadiene)rhodium moiety to one of the arene rings breaks the C_5 symmetry of the corannulene and also renders inequivalent the four vinylic hydrogens of the norbornadiene.

The onset of rotation about the metal-arene axis raises the local symmetry of the norbornadiene to C_2, but the overall molecular symmetry remains as C_1. Migration of the diene-rhodium fragment around the peripheral benzo rings creates a dynamic C_5 symmetry for the corannulene unit but the norbornadiene retains its two-fold character. When the norbornadiene is itself homochiral, as in (R,R)-2,5-dimethyl-bicyclo[2.2.1]hepta-2,5-diene (Scheme 5.22), the result is a dynamic resolution of the equilibrating enantiomers of the *sym*-pentasubstituted corannulenes.[91]

5.8. The Dynamic Behaviour of Triptycenes

The Diels-Alder cycloaddition of appropriately substituted benzynes to anthracenes (Scheme 5.23) provides a convenient route to a wide

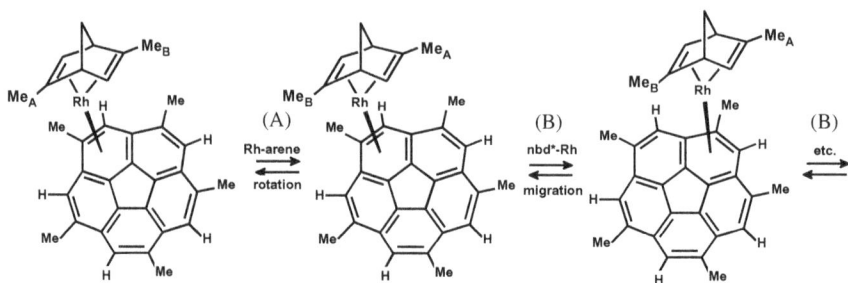

Scheme 5.22. (A) Rotation about the rhodium-arene axis equilibrates the methyls on the norbornadiene; (B) migration of the (nbd*)Rh moiety around the corannulene creates dynamic five-fold symmetry.

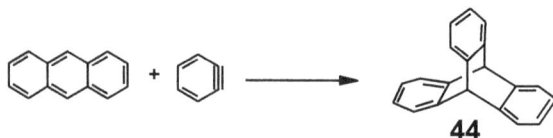

Scheme 5.23. Diels-Alder addition route to triptycene.

variety of triptycenes,[92] such as **44**, a structure comprising three benzene rings in a three-fold symmetric arrangement.[93]

5.8.1. *Triptycenes as Molecular Bevel or Spur Gears*

The triptycene unit, with its triple-bladed paddlewheel architecture, has long attracted the attention of researchers seeking to synthesise nanoscale analogues of gears, brakes and other types of machinery. These studies focussed not merely on the *iconic* aspects whereby the molecule resembled its macroscale counterpart in appearance, but rather as *analogic* models that exhibited the same dynamic behaviour as the larger-scale object.[65] This area was first summarised in a historically important early review describing the results of an exceptionally fruitful Japan-USA collaboration by two world-leading pioneers, Hiizu Iwamura from Nihon University, Tokyo, and Kurt Mislow from Princeton, New Jersey. They described how ditriptycyl species, **45**

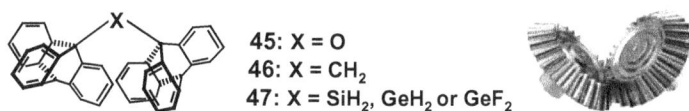

| 45: X = O |
| 46: X = CH$_2$ |
| 47: X = SiH$_2$, GeH$_2$ or GeF$_2$ |

Figure 5.18. Molecular and mechanical bevel gears in which the rotating fragments are oriented at an angle to each other.

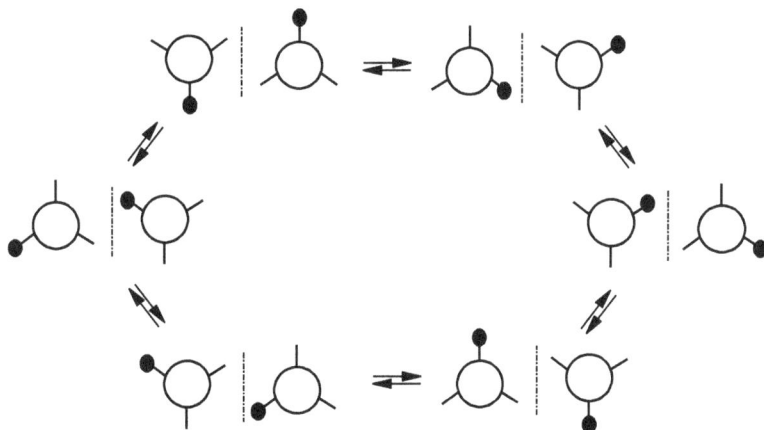

Figure 5.19. Conformational *disrotatory* gearing circuit for the *meso* phase isomers of Tp$_2$X, where the filled circles indicate substituted benzene rings.

and **46**, shown in Figure 5.18, bearing adjacent triptycyl groups connected by an angular linkage (methylene or oxygen) exhibit rapid cogwheeling which requires the interaction of tightly meshed gears.[68]

In such molecules, the barrier to correlated gear disrotation is very low, just one or two kcal mol^{-1}, whereas gear slippage has a remarkably high barrier (~30 kcal mol^{-1}) as was shown by labelling of the triptycyl units. When one benzo ring of each triptycyl bears a single substituent, such as methyl or chloro, the molecule exists as both *meso* and *d,l* stereoisomers. These are designated as *phase isomers* that do not interconvert, and indeed can be separated chromatographically, isolated, and fully characterised, unequivocally demonstrating their independent existence. In the *meso* case, as the molecule undergoes gear rotation, shown in Figure 5.19, the labelled rings exhibit

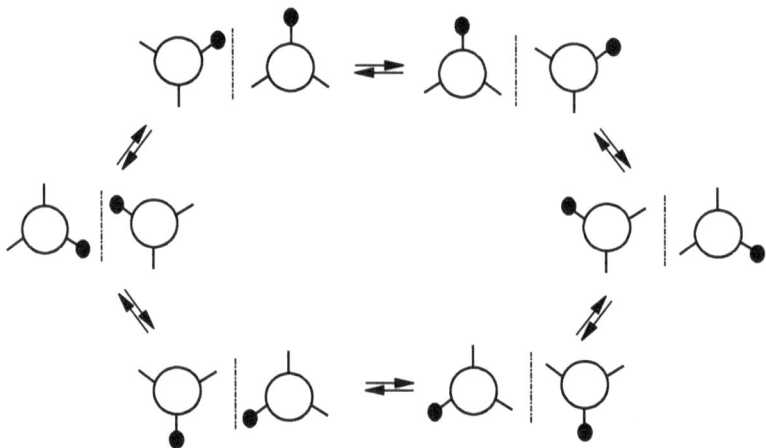

Figure 5.20. Conformational disrotatory gearing circuit for the enantiomeric *d,l* phase isomers of Tp$_2$X, where the filled circles indicate substituted benzene rings.

correlated motion but never come close to each other, and the system exhibits dynamic C_{2v} symmetry.

In contrast, the pair of *d,l*-enantiomers each exhibit dynamic C_2 symmetry. One such conformational circuit is illustrated in Figure 5.20 (the other is simply its mirror image) and reveals that in this case the labelled rings do come into close proximity in the course of the gearing.

More recently, it has been reported that when the ditriptycyl moieties are linked by larger molecular fragments, such as in **47** (SiH$_2$, GeH$_2$ or GeF$_2$), the barrier to gear slippage is reduced to 16–18 kcal mol^{-1},[94] and was attributed to the increased bond lengths and wider angles between the rotating units than was found in the original systems linked by carbon or oxygen bridges.

Rapid molecular cogwheeling resembling the behaviour of spur gears has also been reported; typically, the di-indenyl system, **48** (Figure 5.21), and the bibenzimidazole, **49** (Figure 5.22), in which each bears adjacent triptycyl units, have been characterised spectroscopically and also by X-ray crystallography.[95,96]

In a very impressive display of molecular cogwheeling, the cobalt-catalysed trimerisation of a 1,6-di-(9-triptycyl)hexa-1,3,5-triyne, **50**, yielded the hexa-(9-ethynyltriptycyl)benzene, **51**, which functioned as a sextuple gear (Scheme 5.24). The peripheral triptycyl

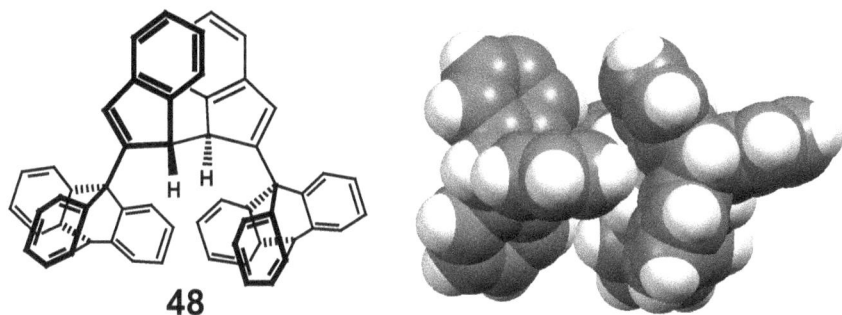

48

Figure 5.21.   Space-filling view of the X-ray crystal structure of **48** showing the tightly meshed triptycyl groups.

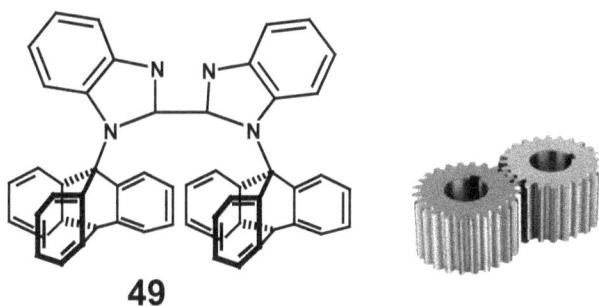

49

Figure 5.22.   Molecular and mechanical spur gears in which the rotational axes are parallel.

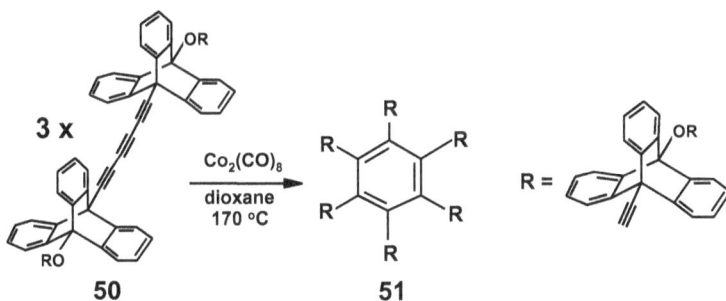

Scheme 5.24.   Trimerisation of a hexa-1,3,5-triyne to form a hexa-substituted benzene.

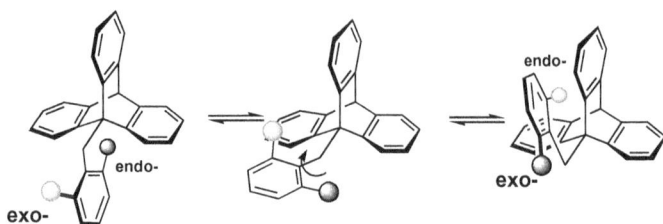

Figure 5.23.   Rotational behaviour of 9-(2,6-dimethylbenzyl)-triptycene, **52**.

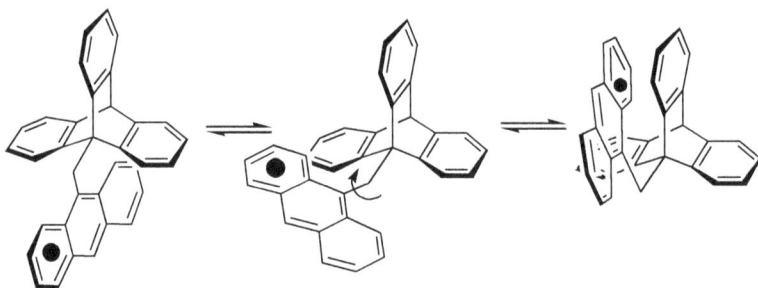

Figure 5.24.   The rotational behaviour of 9-(9-anthracenylmethyl)-triptycene, **53**, demonstrating correlated gear rotation.

moieties were oriented alternately "up" and "down" so as to form a cyclic tightly meshed sextuple gear that rotates freely. However, upon complexation of a (C_5Me_5)Ru fragment onto a benzene ring of one of the triptycyl substituents, rotation is partially blocked, the barrier is markedly enhanced, and may also bring about ring slippage.[97]

In another example of molecular gearing, we consider the dynamic behaviour of 9-(2,6-dimethylbenzyl)triptycene, **52**, in which the aryl substituent is aligned almost periplanar with the C(9)-CH_2 linkage. As illustrated in Figure 5.23, rotation about this bond by 120° is accompanied by rotation about the CH_2–C(Ar) bond by 180°, thus functioning as a molecular scale 2:3 bevel gear system comprised of two- and three-toothed wheels. This process was conveniently followed by monitoring the ^{1}H NMR resonances of the benzyl methyls and the paddlewheel blades of the triptycene unit.[98] Rotation about the three-fold axis of the triptycene equilibrated the three blades and, concomitantly, the *exo* and *endo* methyl groups with an observed barrier of ~10 kcal mol^{-1}.

A perhaps even more impressive example is provided by (9-anthracenyl)(9-triptycyl)methane, **53**, whereby rotation about the two axes linked by the bridging methylene unit equilibrates the paddlewheel blades and also brings about interconversion of the outer benzo rings of the anthracene substituent.[99] Most importantly, in each case the inherent mirror symmetry of the molecule does not differentiate clockwise or counter-clockwise rotation which proceed at equal rates, as expected from the principle of microscopic reversibility.[100]

5.8.2. Ferrocenyl Triptycenes

The development of high-yield routes to ferrocenyltriptycenes prompted an investigation of their dynamic behaviour. Under conditions of rapid rotation, the three paddlewheel blades in 9-ferrocenyltriptycene are NMR equivalent, but are split 2:1 in the low temperature limit when ferrocenyl rotation is slowed on the NMR timescale; the barrier was found to be 17 kcal mol^{-1}. However, 9,10-diferrocenyltriptycene, **54**, was found to exist as co-existing, slowly interconverting *meso* and *d,l* rotamers (Scheme 5.25) in which the ferrocenyl substituents were eclipsed and staggered, respectively.[101]

These isomers are readily differentiated since in the *meso* (C_{2v}) case the mirror-related benzo rings give rise to *two* sets of CH environments, whereas in the C_2-symmetric *rac* isomer one sees *four* such resonances in both the ^{1}H and ^{13}C NMR regimes. The structure of

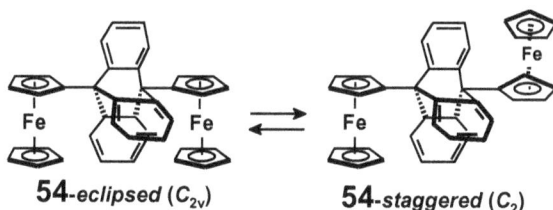

54-*eclipsed* (C_{2v})   **54**-*staggered* (C_2)

Scheme 5.25. Interconversion of *meso*- and *d,l*-9,10-diferrocenyltriptycene.

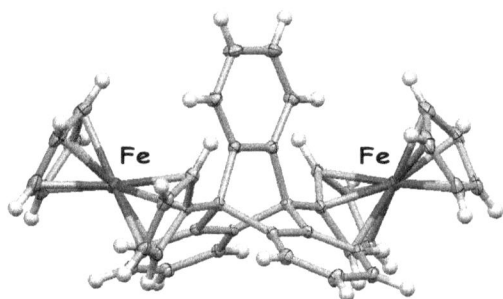

Figure 5.25.   The molecular structure of the C_2-symmetric rotamer of 9,10-diferrocenyltriptycene, **54**.

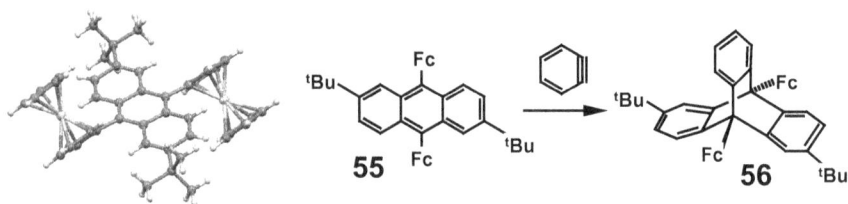

Figure 5.26.   Molecular structure of 2,6-di-*tert*-butyl-9,10-diferrocenylanthracene, **55**, and the synthetic route to 2,6-di-*tert*-butyl-9,10-diferrocenyltriptycene, **56**.

one of the C_2-symmetric *racemic* rotamers, whereby the two ferrocenyls are aligned with different valleys, appears as Figure 5.25. A 2D-EXSY study revealed that the exchange process occurs by successive 120° rotations of a single ferrocenyl substituent, akin to a set of molecular dials. Thus, interconversion of the *d* and *l* rotamers requires two rotations, passing via the *meso* intermediate.[101]

Taking this one stage further, addition of benzyne to 2,6-di-*tert*-butyl-9,10-diferrocenylanthracene, **55**, yields the corresponding triptycene, **56**, in which the three-fold degeneracy of the paddlewheel fragment is broken, and the triptycene is inherently C_2-symmetric (Figure 5.26).[102] Since each ferrocenyl group can be aligned within any one of three inter-blade valleys, this gives rise to nine rotamers, three of which are singly degenerate, while the remaining six exist as three doubly degenerate pairs. Elucidation of the dynamic behaviour of such an apparently complex six-component system might appear to be insoluble, but this turned out not to be the case.

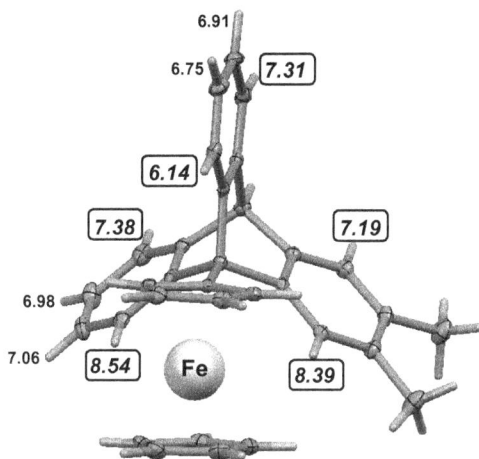

Figure 5.27. ¹H NMR chemical shifts (ppm) of the *ortho* protons in 9-ferrocenyl-2,3-dimethyl triptycene, **57**.

The identification of the individual rotamer subspectra and their rates of interchange was greatly simplified by the remarkably large diamagnetic anisotropy of the ferrocenyl moiety. As shown in Figure 5.27, in 9-ferrocenyl-2,3-dimethyltriptycene, **57**, the spread of chemical shifts between protons aligned directly above the iron atom and those sited close to the plane between the cyclopentadienyl rings containing the metal is ~2.4 ppm.

Hence, on a 600 MHz spectrometer, taking advantage of all the standard two-dimensional NMR techniques, the problem is tractable, the ¹H and ¹³C spectra are fully assignable, and the barriers to all the rotamer exchange processes have been fully evaluated. For example, direct one-step exchange between rotamers **A ↔ E**, or **C ↔ F**, is viable, whereas the interconversion **A ↔ C** is at least a two-step process (Figure 5.28).[102]

5.8.3. *Triptycenes as Components of Organometallic Molecular Brakes*

A particularly ingenious experiment involving the incorporation of a triptycyl unit in a molecular brake has been reported by Ross Kelly from Boston College, a pioneer in the field of molecular machinery. In

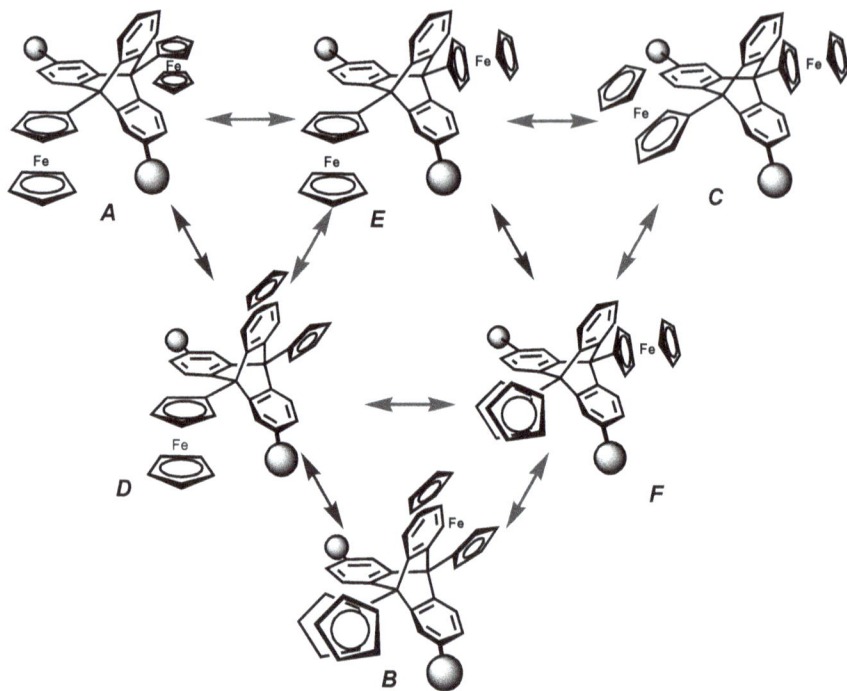

Figure 5.28. Possible one-step interconversion pathways between conformers of 2,6-di-*tert*-butyl- 9,10-diferrocenyltriptycene, **56**.

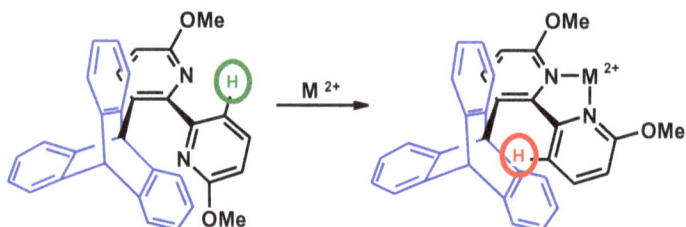

Figure 5.29. Ross Kelly's molecular brake initiated by complexation of a metal ion to the pyridyl nitrogens in **58**.

this case, a bipyridyl substituent was attached at C(9) and, initially, the triptycyl fragment in **58** was free to rotate. However, complexation of both bipyridyl nitrogens to a metal (normally Hg^{2+}) forces a pyridine ring hydrogen in between two of the blades of the paddlewheel and breaks its three-fold symmetry, as shown in Figure 5.29.[103]

An attempt to develop an organometallic molecular brake was described by Andrew Stephens and Chris Richards from the University of Wales in Cardiff who synthesised [(9-triptycylethynyl)cyclopentadienyl](tetraphenylcyclobutadiene)cobalt, **59**. It was hoped that rotation of the paddlewheel would be gated by the bulky C_4Ph_4 ligand; unfortunately, however, there was no unambiguous spectroscopic evidence to support such a scenario.[104]

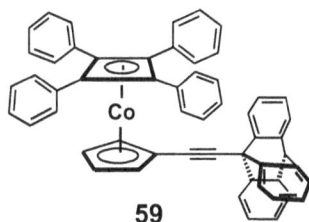

59

A different approach to breaking the three-fold symmetry of the triptycene as a component of a molecular brake was based upon the migration of an organometallic moiety.[105] The haptotropic shift of a metal carbonyl or other organometallic fragment over the surface of an indene is a well-established occurrence.[106] Typically, as shown in Scheme 5.26, deprotonation of the $[(\eta^6\text{-indene})Mn(CO)_3]^+$ cation, **60**, induces migration of the tricarbonylmanganese moiety from the six- to the five-membered ring, as in **61**.

This phenomenon was exploited to bring about the reversible η^6- to η^5-haptotropic shift of a tricarbonylchromium moiety across an indenyl framework, from **62** to **63** (Scheme 5.27), so as to block the rotation of the triptycyl paddlewheel and thus activate the first

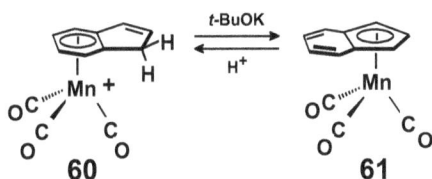

Scheme 5.26. The base-promoted haptotropic shift of $[(\eta^6\text{-indene})Mn(CO)_3]^+$, **60**, to $(\eta^5\text{-indenyl})Mn(CO)_3$, **61**.

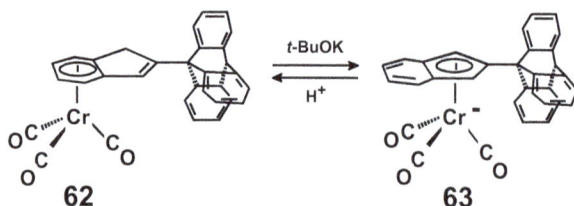

Scheme 5.27. The triptycyl paddlewheel in **62** is free to rotate but, after migration of the metal tricarbonyl, rotation is blocked in **63**.

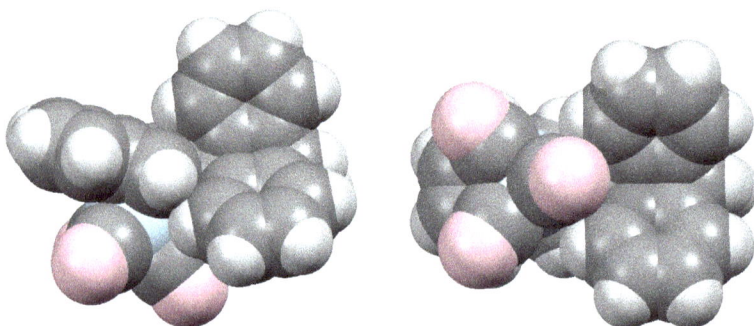

Figure 5.30. Molecular structure of $[\eta^5\text{-}2\text{-}(9\text{-triptycyl})$indenyl]tricarbonyl-manganese(I), as for **63**, but M = Mn; the space-fill representations are viewed from the side, and from below, clearly revealing the mirror symmetry of the molecule.

organometallic molecular brake. The migration of the bulky organo-metallic group over a distance of approximately 2 Å breaks the three-fold symmetry of the triptycyl unit and renders two of the blades no longer equivalent to the third, as clearly indicated in the ^{1}H and ^{13}C NMR spectra of **63**. Analogous compounds of $Mn(CO)_3$ and $Re(CO)_3$ were also prepared; as seen in the chromium system, the NMR data again exhibited 2:1 splitting of the paddlewheel blades in the manganese and rhenium η^5-complexes.[107]

The structure of the molecular brake was independently established via X-ray crystallography that unequivocally revealed the positioning of the metal tripod such that a carbonyl ligand is tightly embedded within a valley between two of the triptycene blades. This is illustrated in Figure 5.30 which presents space-fill views both from the side and from below the paddlewheel.[107]

5.9. Dynamic Behaviour of Three-bladed Propellors

As a final example, we present the remarkable story of the elucidation of the enantiomerisation mechanism of some three-bladed propellors of the type Ar$_3$Z (e.g., triarylboranes) and Ar$_3$ZX (such as triarylmethanes or triarylphosphine oxides). The complete resolution of this problem involved a combination of NMR spectroscopy and the Bürgi-Dunitz X-ray crystallographic approach.

5.9.1. *NMR Studies of Triarylboranes or Triarylmethanes*

Molecules Ar$_3$B or Ar$_3$CH exist as three-fold symmetric propellors of D_3 or C_3 symmetry, respectively, that undergo stereoisomerisation so as to interconvert their enantiomers (or diastereomers, depending on their substitution pattern). One can envisage a number of mechanistic scenarios, but one possibility, the simultaneous conrotatory motion of all three aryl substituents, could be readily discounted since it would bring about serious steric hindrance in the transition state. Instead, as explicated by Kurt Mislow, undoubtedly the most influential stereochemist in recent decades, and his group from Princeton University, this rearrangement occurs by a *two-ring flip* process whereby, when viewed from the central atom, as one blade rotates in one direction (say clockwise) the other two blades both rotate counter-clockwise and eventually bring about chirality inversion. The rearrangement process, viewed along the three-fold axis, is shown in Scheme 5.28.

This proposal was verified not only by Empirical Force Field calculations but more directly by carefully designed variable-temperature NMR measurements. Full details of these studies are provided in a

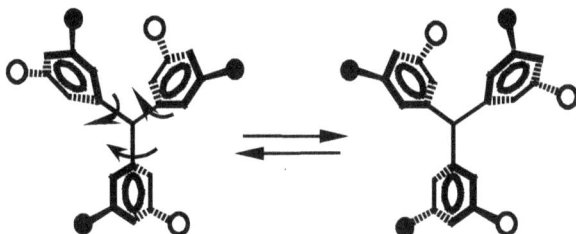

Scheme 5.28. The two-ring flip mechanism for chirality inversion in Ar$_3$ZX systems.

Mislow review,[108] but essentially they depend on the existence of *phase isomers* (or as it was originally designated, *residual stereoisomerism*), a phenomenon already discussed in Section 5.8.1. These revealed that the correlated behaviour of these tightly gear-meshed molecular propellors undergoing a series of two-ring flips allows chirality inversion while maintaining the stereochemical integrity of the phase isomers.

5.9.2. *The Bürgi-Dunitz Structure-correlation Approach*

Prior to addressing the question of chirality inversion in three-bladed propellors, we should remind ourselves of the Bürgi-Dunitz approach to the determination of chemical reaction paths. Hans Beat Bürgi (University of Bern) and Jack Dunitz (ETH Zürich) were very distinguished X-ray crystallographers who developed the idea of extracting data from the Cambridge Structural Database (CSD) on a large number of closely related structures, and then using it to establish reaction pathways experimentally.[109]

The classic example is their determination of the trajectory of attack by a nucleophile at a carbonyl center from a very large number of closely related structures in which the position of the nucleophile changes relative to the carbonyl group that is undergoing a gradual transition from trigonal planar to tetrahedral.[110] This approach trajectory, now known as the Bürgi-Dunitz angle, is ~107°, close to the tetrahedral angle; as with many important discoveries, with 20/20 hindsight this result is perhaps obvious, but was only revealed as the result of their remarkable insight. This approach has since been applied in many other cases, such as the migration of a terminal carbonyl ligand between metal centres via a bridging carbonyl intermediate,[111] or the racemisation of the tricobalt cluster cation $[(OC)_9Co_3CCH_2]^+$, **11**,[112] as discussed in Section 5.4.2.

5.9.3. *An X-ray Crystallographic Study of Triarylphosphine Oxides and Related Molecules*

The Bürgi-Dunitz approach to the question of the mechanism of stereoisomerisation of three-bladed molecular propellors perfectly

Figure 5.31. Views of the molecular geometry along the stereoisomerisation pathway. (Left) C_3-symmetric with torsion angles 40°, 40°, 40°; (centre) mirror-symmetric with angles 90°, –10°, 10°; (right) enantiomer of the first structure with angles –40°, –40°, –40°.

complements the two-ring flip process derived from Mislow's NMR study. In the X-ray structures of **560** molecules of the type Ph_3P–X, where X = O, N and many other elements, the structures of triphenylphosphine moieties were found to range from an almost idealised C_3 geometry, when viewed down the P=O axis, towards a mirror-symmetric arrangement depicted in Figure 5.31. The equilibrium structure of an isolated Ph_3P–X fragment is a three-fold symmetric propellor in which all the phenyl rings are rotated in the same sense with a torsion angle of ~**40°** with respect to the C–P–X plane. However, as one examines the less populated regions of the distribution of structures, the stereoisomerisation pathway becomes more apparent.[113]

From the viewpoint of an observer situated on the central atom, starting from a molecule with positive torsion angles of 40°, as ring *A* rotates clockwise toward **90°** (i.e., perpendicular to the original C_3 axis) its motion is coupled with counter-clockwise rotation of the other two rings, which gradually get out of step. Rings *B* and *C* rotate through –30° and –50° so as to reach a mirror-symmetric structure in which the *A*, *B*, *C* torsion angles are now 90°, –10° and 10°, respectively. The rearrangement then continues, leading eventually to the propellor with inverted chirality (all torsion angles now –40°).

We note in passing that molecules of the type Tp_3XZ, for example where X = Ge and Z = Cl, are also known.[114] Such systems exhibit structural rigidity as the result of the odd number of securely interlocked triptycyl rotors. The transmission of cooperative torsional motion along macromolecular chains of suitable meshed *N*-chemical gears is governed by the parity rule for gear chains, which dictates correlated disrotation of the terminal gears when N is even, and conrotatory interactions when N is odd. Thus, arranging an odd number of gears in a cyclic array leads to frustration of the cogwheeling process.

5.10. Concluding Remarks

Broken symmetry is at the core of many physical and chemical phenomena, ranging from the generation of mass by the Higgs boson, to the predominance of matter over anti-matter, to the chiral nature of so many natural products, and the right-handed character of the DNA double helix. Applying these concepts to molecular stereochemistry and dynamics, we have seen how judiciously broken symmetry can provide information about the mechanisms and energetics of otherwise undetectable rearrangement processes.

Evidently, one can only provide representative examples of some of the many ways ingenious chemists have tackled this problem. These include such methods as (a) incorporation of a suitable isotope such as deuterium, carbon-14, nitrogen-15, iron-59, etc., (b) replacement of one molecular fragment by another, e.g., CO by CS or NO, diphos by arphos, phenyl by naphthyl, $(C_5H_5)Ni$ by $Co(CO)_3$, etc., (c) introduction of a diastereotopic unit, such as isopropyl, to probe the chirality of the system, (d) addition of a substituent to break mirror symmetry, e.g., a 1-methyl group in the indenyl framework, or a π-bonded organometallic fragment to discriminate between faces of a planar cyclic system, and (e) monitoring the dynamic behaviour of a system containing multiple stereocentres.

In recent decades, advances in NMR technology have given us immensely higher field strengths, a plethora of new pulse sequences, multi-dimensional spectra, etc., but classical variable-temperature measurements are still a valuable component of our arsenal of investigative techniques. One should, however, not undervalue the contributions of the synthetic chemist in this regard; it is relatively easy to draw a hypothetical molecule with substituents appropriate for an NMR study, but more of a challenge to devise a synthetic route and to prepare an analytically pure sample. Nevertheless, we can be confident that symmetry breaking will endure as an important technique for our understanding of complex molecular dynamics, and we look forward with anticipation to even more elegant revelations in the coming years.

References

1. T. Helgaker, M. Jasuński, P. Garbacz and K. Jackowski, The NMR indirect nuclear spin-spin coupling constant of the HD molecule. *Mol. Phys.* **2012**, *110*, 2611–2617.

2. J.J. Drysdale and W.D. Phillips, Restricted rotation in substituted ethanes as evidenced by nuclear magnetic resonance. *J. Am. Chem. Soc.* **1957**, *79*, 319–322.

3. K. Mislow and M. Raban, Stereochemical relationships of groups in molecules. In *Topics in Stereochemistry*, Interscience: New York, USA, 1967; Vol. 1, pp. 1–138.

4. V.I. Sokolov, P.V. Petrovskii and O.A. Reurov, Chirality in α-ferrocenycarbonium ion as monitored using nuclei diastereotopism. *J. Organomet. Chem.* **1973**, *59*, C27–C29.

5. R. Gleiter, R. Seeger, H. Binder, E. Fluck and M. Cais, Partial charge on iron in diferrocenymethylium ion. *Angew. Chem. Int. Ed.* **1972**, *11*, 1028–1029.

6. M.J. McGlinchey, Ferrocenyl migrations and molecular rearrangements: the significance of electronic charge delocalization. *Inorganics* **2020**, *8*, 68.

7. J. Reny, C.Y. Wang, C.H. Bushweller and W.G. Anderson, Stereodynamics of N-isopropyl-N,N-dialkylamines. Direct NMR observation of diastereotopic isopropyl methyl groups and slow nitrogen inversion. *Tetrahedron Lett.* **1975**, *16*, 503–506.

8. J.H. Brown and C.H. Bushweller, Stereodynamics of isopropyldimethylamine. $^{13}C\{^1H\}$ and ^1H dynamic NMR studies. Molecular mechanics calculations. *J. Am. Chem. Soc.* **1992**, *114*, 8153–8158.

9. E. Fluck and K. Isslieb, NMR investigations of phosphorus compounds, VIII. ^{31}P spectra of phosphines, diphosphines and diphosphinedisulfides. *Chem. Ber.* **1965**, *98*, 2674–2680.

10. J.B. Lambert and D.C. Mueller, The inversion of diphosphines. *J. Am. Chem. Soc.* **1966**, *88*, 3669–3670.

11. J.B. Lambert, G.F. Jackson III and D.C. Mueller, The stereochemical lability of diphosphines and diarsines. *J. Am. Chem. Soc.* **1968**, *90*, 6401–6405.

12. D.G. Gilheany, No d orbitals but Walsh diagrams and maybe banana bonds: chemical bonding in phosphines, phosphine oxides and phosphonium ylides. *Chem. Rev.* **1994**, *94*, 1339–1374.

13. D.R. Rayner, A.J. Gordon and K. Mislow, Thermal racemization of diaryl, alkyl, aryl, and dialkyl sulfoxides by pyramidal inversion. *J. Am. Chem. Soc.* **1968**, *90*, 4854–4860.

14. M. Kainosho, K. Ajisaka, W.H. Pirkle and S.D. Beare, Use of chiral solvents or lanthanide shift reagents to distinguish *meso* from *d* or *l* diastereomers. *J. Am. Chem. Soc.* **1972**, *94*, 5924–26.

15. J-P. Albrand and J-B. Robert, Assignment of *meso* and (±) diastereomers of diphosphines using ^{31}P nuclear magnetic resonance spectroscopy in a chiral solvent. *J.C.S. Chem. Commun.* **1976**, 876–877.

16. J.C. Richards and I.D. Spenser, ^{2}H NMR spectroscopy as a probe of the stereochemistry of enzymic reactions at prochiral centres. *Tetrahedron* **1983**, *39*, 3549–3568.

17. B.D. Allen, J-C. Cintrat, N. Faucher, P. Berthault, B. Rousseau and D.J. O'Leary, An isosparteine derivative for stereochemical assignment of stereogenic (chiral) methyl groups using tritium NMR: theory and experiment. *J. Am. Chem. Soc.* **2005**, *127*, 412–420.

18. A. Salomone, F.M. Perna, A. Falcicchio, S.O. Nilsson Lill, A. Moliterni, R. Michel, S. Florio, S. Stalke and V. Capriati, Direct observation of a lithiated oxirane: a synergistic study using spectroscopic, crystallographic, and theoretic methods on the structure and stereodynamics of lithiated *ortho*-trifluoromethyl styrene oxide. *Chem. Sci.* **2014**, *5*, 528–538.

19. I. Alkorta, C. Dardonville and J. Elguero, Observation of diastereotopic signals in ^{15}N NMR spectroscopy. *Angew. Chem. Int. Ed.* **2015**, *54*, 3997–4000.

20. T.A. Powers and S.A. Evans Jr., Oxygen-17 nuclear magnetic resonance spectroscopy of organosulfur compounds. Part III. ^{17}O NMR lanthanide-induced shifts (LIS) of diastereotopic oxygen atoms in *trans*-2-[alkyl(aryl)sulfonyl] cyclohexanols. *Tetrahedron Lett.* **1990**, *31*, 5835–5838.

21. D.A. Kleier and G. Binsch, Quantum Chemistry Program Exchange, 1969. *A widely used programme initially, that has since been significantly upgraded.*

22. R.E.D. McClung EXCHANGE – Program for the simulation of NMR spectra of exchanging systems, 1990. *An excellent programme for the simulation of* ^{13}C *spectra involving multiple sites, such as in metal carbonyl exchange processes.*

23. H.S. Gutowsky and C.H. Holm, Rate processes and nuclear magnetic resonance spectra. II. Hindered internal rotation of amides. *J. Chem. Phys.* **1956**, *25*, 1228–1234.

24. K. Nikitin and R. O'Gara, Mechanisms and beyond: elucidation of fluxional dynamics by exchange NMR spectroscopy. *Chem. Eur. J.* **2019**, *25*, 4551–4589.

25. B. Jurado and C.S. Springer Jr., Direct measurement of enantiomerization of labile aluminium(III) β-diketonates. *Chem. Comm.* **1971**, 85–87.

26. N. Serpone and D.G. Bickley, Kinetics and mechanisms of isomerization and racemization processes of six-coordinate chelate complexes. *Prog. Inorg. Chem.* **1972**, *17*, 391–566.

27. N. Serpone and D.G. Bickley, Configurational rearrangements in cis-M(AA)$_2$X$_2$, cis M(AA)$_2$XY, and cis-M(AB)$_2$X$_2$ complexes 10. The cis-M(AB)$_2$X$_2$ system – diastereotopic probe on the X-ligands. *Inorg. Chim. Acta.* **1982**, *57*, 211–216, and references therein.

28. M.J. McGlinchey, L. Girard and R. Ruffolo, Cluster-stabilized cations: syntheses, structures, molecular dynamics and reactivity. *Coord. Chem. Rev.* **1995**, *143*, 331–381.

29. B.E.R. Schilling and R. Hoffmann, M$_3$L$_9$(ligand) complexes. *J. Am. Chem. Soc.* **1979**, *101*, 3456–3467.

30. R.T. Edidin, J.R. Norton and K. Mislow, Evidence for tilted ground-state structures and fluxionality in Co$_3$(CO)$_9$CCHR$^+$. *Organometallics* **1982**, *1*, 561–562.

31. K.A. Sutin, J.W. Kolis, M. Mlekuz, P. Bougeard, B.G. Sayer, M.A. Quilliam, R. Faggiani, C.J.L. Lock, M.J. McGlinchey and G. Jaouen, Arphos and diphos complexes of Co$_2$(CO)$_6$MC-CO$_2$-i-Pr [M = Co(CO)$_3$, (C$_5$Me$_5$)Mo(CO)$_2$]: X-ray crystal structure and NMR fluxionality. *Organometallics* **1987**, *6*, 439-447.

32. H. Vahrhenkamp, Framework chirality and optical activity of organometallic cluster compounds. *J. Organomet. Chem.* **1989**, *370*, 65–73.

33. H. El Hafa, C. Cordier, M. Gruselle, Y. Besace, G. Jaouen and M.J. McGlinchey, An NMR study of the dynamic behavior of [(2-propynylbornyl)-Mo$_2$(CO)$_4$Cp$_2$]$^+$BF$_4^-$: non-fluxional Mo-Co clusters as the key to understanding the mechanism of the formation of metal-stabilized cations. *Organometallics* **1994**, *13*, 5149–5156.

34. D.T. Clark, K.A. Sutin and M.J. McGlinchey, Heterometallic alkylidyne clusters: natural products as sources of chirality. *Organometallics* **1989**, *8*, 155–161.

35. D. Shriver, H.D. Kaesz and R.D. Adams, Eds. *The Chemistry of Metal Cluster Complexes*; Wiley-VCH: New York, USA, 1990.

36. M. Mlekuz, P. Bougeard, B.G. Sayer, M.J. McGlinchey, S. Peng, A. Marinetti, J-Y. Saillard, J. Ben Naceur, B. Mentzen and G. Jaouen, Syntheses, crystal structures and DNMR studies on the mixed clusters CpNiFe(CO)$_3$(RC≡CR') M, (M = CpNi, Co(CO)$_3$, Mo(CO)$_2$Cp): some comments on the acetylene rotation process. *Organometallics* **1985**, *4*, 1123–1130.

37. B.E.R. Schilling and R. Hoffmann, Dependence of equilibrium geometry and rearrangement modes on electron count in one class of trinuclear complexes of acetylene. *Acta Chem. Scand., Ser. B* **1979**, *B33*, 231–232.

38. R. Hoffmann, Building bridges between inorganic and organic chemistry. *Angew. Chem. Int. Ed. Engl.* **1982**, *21*, 711–724.

39. W-D. Stohrer and R. Hoffmann, Bond stretch isomerism and polytopal rearrangements in (CH)$_5^+$, (CH)$_5^-$ and (CH)$_4$CO. *J. Am. Chem. Soc.* **1972**, *94*, 1661–1668.

40. W.E. Barth and R.G. Lawton, The synthesis of corannulene. *J. Am. Chem. Soc.* **1971**, *93*, 1730–1745.

41. L.T. Scott, M.M. Hashemi, D.T. Meyer and H.B. Warren, Corannulene. A convenient new synthesis. *J. Am. Chem. Soc.* **1991**, *113*, 7082–7083.

42. L.T. Scott, P-C. Cheng, M.M. Hashemi, M.S. Bratcher, M.T. Meyer and H.B. Warren, Corannulene. A three-step synthesis. *J. Am. Chem. Soc.* **1997**, *119*, 10963–10968.

43. A. Borchardt, A. Fichicello, K.V. Kilway, K.K. Baldridge and J.S. Siegel, Synthesis and dynamics of the corannulene nucleus. *J. Am. Chem. Soc.* **1992**, *114*, 1921–1923.

44. A.M. Butterfield, B. Gilomen and J.S. Siegel, Kilogram-scale production of corannulene. *Org. Process Res. Dev.* **2012**, *16*, 664–676.

45. Y-T. Wu and J.S. Siegel, Aromatic molecular bowl hydrocarbons: synthetic derivatives, their structures, physical properties. *Chem. Rev.* **2006**, *106*, 4843–4867.

46. L.T. Scott, M.M. Hashemi and M.S. Bratcher, Corannulene bowl-to-bowl inversion is rapid at room temperature. *J. Am. Chem. Soc.* **1992**, *114*, 1920–1921.

47. P.U. Biedermann, S. Pogodin and I. Agranat, Inversion barrier for corannulene. A benchmark for bowl-to-bowl inversions in fullerene fragments. *J. Org. Chem.* **1999**, *64*, 3655–3662.

48. L.A.F. Andrade, L.A. Zeoly, R.A. Cormanich and M.P. Freitas, Conformational signature of Ishikawa's reagent using NMR information from diastereotopic fluorines. *Beilstein J. Org. Chem.* **2019**, *15*, 506–512.

49. M. Karplus, Vicinal proton coupling in nuclear magnetic resonance. *J. Am. Chem. Soc.* **1963**, *85*, 2870–2871.

50. S.J. Pike, M. De Poli, W. Zawodny, J. Raftery, S.J. Web and J. Clayden, Diastereotopic fluorine substituents as ^{19}F NMR probes of screw-sense preference in helical foldamers. *Org. Biomol. Chem.* **2013**, *11*, 3168–3176.

51. S. Meyer, J. Häfliger, M. Schäfer, J.J. Molloy, C.G. Daniliuc and R. Gilmour, A chiral pentafluorinated isopropyl group via iodine(I)/(III) catalysis. *Angew. Chem. Int. Ed.* **2021**, *60*, 6430–6434.

52. M.J. McGlinchey, Hexaethylbenzene: a sterically crowded arene and conformationally versatile ligand. *ChemPlusChem* **2018**, *83*, 480–499.

53. G. Hunter, D.J. Iverson, K. Mislow and J.F. Blount, Conformational variability in hexaethylbenzene π-complexes. Crystal and molecular structure of tricarbonyl(hexaethylbenzene)chromium(0) and dicarbonyl(hexaethylbenzene)(triphenylphosphine)chromium(0). *J. Am. Chem. Soc.* **1978**, *100*, 6902-6904.

54. M.M. Maricq, J.S. Waugh, J.L. Fletcher and M.J. McGlinchey, Anisotropic ring-carbon shifts in arene chromium tricarbonyl complexes. *J. Am. Chem. Soc.* **1978**, *100*, 6902-6904.

55. M.J. McGlinchey, J.L. Fletcher, B.G. Sayer, P. Bougeard, R. Faggiani, C.J.L. Lock, A.D. Bain, C.A. Rodger, E.P. Kündig, D. Astruc, J-P. Hamon, P. LeMaux, S. Top and G. Jaouen, Stopping a chromium carousel: X-ray crystallographic and

variable-temperature ^{13}C NMR studies on dicarbonyl(hexaethylbenzene)-thiocarbonylchromium(0) and related complexes. *Chem. Commun.* **1983**, 634-636.

56. M.J. McGlinchey, P. Bougeard, B.G. Sayer, R. Hofer and C.J.L. Lock, Solid state ^{13}C and high resolution ^{1}H and ^{13}C NMR spectra of dicarbonyl(hexaethylbenzene)thiocarbonylchromium(0): a reaffirmation of slowed tripodal rotation. *Chem. Commun.* **1984**, 789–790.

57. B. Mailvaganam, C.S. Frampton, B.G. Sayer, S. Top and M.J. McGlinchey, An X-ray crystallographic and high field NMR study of $[(C_6Et_6)Cr(CO)_2NO]^+$ BF_4^- and of $[C_6Et_6)Cr(CO)(CS)NO]^+$ BF_4^-: steric inhibition of tripodal rotation. *J. Am. Chem. Soc.* **1991**, *113*, 1177–1185.

58. M.J. McGlinchey, Slowed tripodal rotation in arene-chromium complexes: Steric and electronic barriers. *Adv. Organomet. Chem.* **1992**, *34*, 285–325.

59. J.A. Chudek, G. Hunter, R.L. MacKay, P. Kremminger, W. Weissensteiner, Restricted rotation about a metal-arene bond caused by the syteric effects of proximal ethyl groups; stereodynamics of some complexes of 1,3,5-triethyl-2,4,6-tris(trimethylsilyl)benzene. *J. Chem. Soc., Dalton Trans.* **1991**, 3337–3347.

60. P.A. Downton, B. Mailvaganam, C.S. Frampton, B.G. Sayer and M.J. McGlinchey, Unequivocal proof of slowed chromium tricarbonyl rotation in a sterically crowded arene complex: an X-ray crystallographic and variable-temperature high field NMR study of $(C_6Et_5COCH_3)Cr(CO)_3$. *J. Am. Chem. Soc.* **1990**, *112*, 27-32.

61. K.V. Kilway and J.S. Siegel, Evidence for gated stereodynamics in [1,4-bis(4,4-dimethyl-3-oxopentyl)-2,3,5,6-tetraethylbenzene)chromium tricarbonyl. *J. Am. Chem. Soc.* **1991**, *113*, 2332–2333.

62. J. Siegel, A. Gutiérrez, W.B. Schweizer, O. Ermer and K. Mislow, Static and dynamic stereochemistry of hexaisopropylbenzene: a gear-meshed hydrocarbon of exceptional rigidity. *J. Am. Chem. Soc.* **1986**, *108*, 1569–1575.

63. B. Kahr, S.E. Biali, W. Schaefer, A.B. Buda and K. Mislow, Molecular structure of hexakis(dichloromethyl)benzene. *J. Org. Chem.* **1987**, *52*, 3713–3717.

64. B. Gloaguen and D. Astruc, Chiral pentaisopropylcyclopentadienyl and pentakis(1-ethylpropyl)cyclopentadienyl complexes: one-pot synthesis by formation of 10 carbon-carbon bonds from pentamethylcobalticinium. *J. Am. Chem. Soc.* **1990**, *112*, 4607–4609.

65. S. Brydges, L.E. Harrington and M.J. McGlinchey, Sterically hindered organometallics: multi-n-rotor (n = 5, 6 and 7) and the search for correlated rotations. *Coord. Chem. Rev.* **2002**, *233–234*, 75–105.

66. S. Brydges and M.J. McGlinchey, Conformations and threshold rotational mechanisms of C_5Ar_5 and C_5Ar_4X propellers: a structure correlation and computational study. *J. Org. Chem.* **2002**, *67*, 7688–7698.

67. S. Brydges, J.F. Britten, L.C.F. Chao, H.K. Gupta, M.J. McGlinchey and D.L. Pole, The structure of a seven-bladed propeller: $C_7Ph_7^+$ is not planar. *Chem. Eur. J.* **1998**, *4*, 1199–1203.

68. H. Iwamura and K. Mislow, Stereochemical consequences of dynamic gearing. *Acc. Chem. Res.* **1988**, *21*, 175–182.

69. D. Gust and A. Patton, Dynamic stereochemistry of hexaarylbenzenes. *J. Am. Chem. Soc.* **1978**, *100*, 8175–8181.

70. L.E. Harrington, J.F. Britten, K. Nikitin and M.J. McGlinchey, A synthetic, X-ray, NMR and DFT study of β-naphthil dihydrazone, di(β-naphthyl)acetylene, tetra(β-naphthyl)cyclopentadienone, and hexa(β-naphthyl)benzene: $C_6(C_{10}H_7)_6$ is a disordered molecular propeller. *ChemPlusChem* **2017**, *82*, 433–441.

71. B. Mailvaganam, B.G. Sayer and M.J. McGlinchey, The fluxional behaviour of hexaphenylbenzene chromium tricarbonyl: a variable-temperature ^{13}C NMR study. *J. Organomet. Chem.* **1990**, *395*, 177–185.

72. B. Mailvaganam, B.E. McCarry, B.G. Sayer, R.F. Perrier, R. Faggiani and M.J. McGlinchey, Tricarbonylchromium(0) complexes of 1,3,5-triphenylbenzene: an X-ray crystallographic and high field NMR spectroscopic study. *J. Organomet. Chem.* **1987**, *335*, 213–227.

73. H.K. Gupta, N. Reginato, F.O. Ogini, S. Brydges and M.J. McGlinchey, $SiCl_4$-ethanol as a trimerization agent for organometallics: convenient syntheses of the symmetrically substituted arenes $1,3,5$-$C_6H_3R_3$ where R = (C_5H_4) $Mn(CO)_3$ and $(C_5H_4)Fe(C_5H_5)$. *Can. J. Chem.* **2002**, *80*, 1546–1554.

74. F.O. Ogini, Y. Ortin, A.H. Mahmoudkhani, A.F. Cozzolino, M.J. McGlinchey and I. Vargas-Baca, An investigation of the formation of 1,3,5-heterosubstituted benzene rings by cyclo-condensation of acetyl-substituted organometallic complexes. *J. Organomet. Chem.* **2008**, *693*, 1957–1967.

75. H.K. Gupta, S. Brydges and M.J. McGlinchey, Diels-Alder reactions of 3-ferrocenyl-2,4,5-triphenylcyclopentadienone: Syntheses and structures of the sterically crowded systems C_6Ph_5Fc, C_7Ph_6FcH and $[C_7Ph_6FcH][SbCl_6]$. *Organometallics* **1999**, *18*, 115–122.

76. L.E. Harrington, J.F. Britten and M.J. McGlinchey, Ferrocenyl-penta(β-naphthyl)benzene: Synthesis, structure and molecular dynamics. *Can. J. Chem.* **2003**, *81*, 1180–1186.

77. Y. Yu, A.D. Bond, P.W. Leonard, U.J. Lorenz, T.V. Timofeeva, K.P.C. Vollhardt, G.D. Whitener and A.A. Yakovenko, Hexaferrocenylbenzene. *Chem. Commun.* **2006**, 2572–2574.

78. L.D. Field, T. He, P. Humphrey, A.F. Masters and P. Turner, Manganese complexes of the pentaphenylcyclopentadienyl ligand. *Polyhedron* **2006**, *25*, 1498–1506.

79. L.D. Field, C.M. Lindall, A.F. Masters and G.K.B. Clentsmith, Pentaarylcyclopentadienyl complexes. *Coord. Chem. Rev.* **2011**, *255*, 1733–1790.

80. L. Li., A. Decken, B.G. Sayer, M.J. McGlinchey, P. Brégaint, J.-Y. Thépot, J.-R. Hamon and C. Lapinte, Multiple fluxional processes in chiral molecules: the barriers to aryl and tripodal rotation in $(C_5Ph_5)Fe(CO)(PhMe_2P)CHO$ and $(C_5Ph_5)Fe(CO)_2Br$. *Organometallics* **1994**, *13*, 682–689.

81. H.K. Gupta, M. Stradiotto, D.W. Hughes and M.J. McGlinchey, Reactions of C_6F_5Li with tetracyclone and 3-ferrocenyl-2,4,5-triphenylcyclopentadienone: A ^{19}F NMR and X-ray crystallographic study of hindered pentafluorophenyl rotations. *J. Org. Chem.* **2000**, *65*, 3652–3658.

82. D. Nori-shargha, S. Asadzadeha, F-R. Ghanizadehb, F. Deyhimic, M.M. Aminic and S. Jameh-Bozorghi, Ab initio study of the structures and dynamic stereochemistry of biaryls. *J. Mol. Struct. THEOCHEM* **2005**, *717*, 41–51.

83. W. Nowak and M. Wierzbowska, A theoretical study of geometry and transition moment directions of flexible fluorescent probes – acetoxy derivatives of phenylanthracene. *J. Mol. Struct. THEOCHEM* **1996**, *368*, 223–234.

84. K. Nikitin, Y. Ortin, H. Müller-Bunz and M.J. McGlinchey, Restricted rotation in phenyl-anthracenes: a prediction fulfilled. *Org. Lett.* **2011**, *13*, 256–259.

85. R. Cramer, Olefin coordination compounds of rhodium. The barrier to rotation of coordinated ethylene and the mechanism of olefin exchange. *J. Am. Chem. Soc.* **1964**, *86*, 217–222.

86. R. Cramer, J.B. Kline and J.D. Roberts, Bond character and conformation equilibration of ethylene- and tetrafluoroethylene-rhodium complexes from nuclear magnetic resonance spectra. *J. Am. Chem. Soc.* **1969**, *91*, 2519–2524.

87. P. Caddy, M. Green, E. O'Brien, L.E. Smart and P. Woodward, Reactions of coordinated ligands. Part 22. The reactivity of bis-(ethylene)(indenyl)rhodium in displacement reactions with olefins, dienes and acetylenes. *J. Chem. Soc., Dalton Trans.* **1980**, 962–972.

88. H. Eshtiagh-Hosseini and J.F. Nixon, Ethylene displacement reactions of indenylbis(ethylene)rhodium. *J. Less-Common Met.* **1978**, *61*, 107–121.

89. M. Mlekuz, P. Bougeard, B.G. Sayer, M.J. McGlinchey, C.A. Rodger, M.R. Churchill, J.W. Ziller, S-K. Kang, and T.A. Albright, X-ray crystal structure and molecular dynamics of (indenyl)bis(ethylene)rhodium(I): 500 MHz NMR spectra and EHMO calculations. *Organometallics* **1986**, *5*, 1656-1663.

90. T.B. Marder, J.C. Calabrese, D.C. Roe and T.H. Tulip, The slip-fold distortion of π-bound indenyl ligands – dynamic NMR and X-ray crystallographic studies of (η-indenyl)RhL$_2$ complexes. *Organometallics* **1987**, *6*, 2012–2014.

91. D. Bandera, K.K. Baldridge, A. Linden, R. Dorta and J.S. Siegel, Stereoselective coordination of C_5-symmetric corannulene derivatives with an enantiomerically pure [Rh(nbd*)] metal complex. *Angew. Chem. Int. Ed.* **2011**, *50*, 865–867.

92. G. Wittig and R. Ludwig, Triptycene from anthracene and dehydrobenzene. *Angew. Chem.* **1956**, *68*, 40.

93. V.R. Skvarchenko, V.K. Shalaev and E.I. Klabunovskii, Advances in the chemistry of triptycene. *Russ. Chem. Rev.* **1974**, *43*, 951–966.

94. K. Okamura, Y. Inagaki, H. Momma, E. Kwon and W. Setaka, Gear slippage in molecular bevel gears bridged with a Group 14 element. *J. Org. Chem.* **2019**, *84*, 14636–14643.

95. K. Nikitin, H. Müller-Bunz, Y. Ortin, W. Risse and M.J. McGlinchey, Twin triptycyl spinning tops: a simple case of molecular gearing with dynamic C_2 symmetry. *Eur. J. Org. Chem.* **2008**, 3079–3084.

96. D.K. Franz, A. Linden, K.K. Baldridge and J.S. Siegel, Molecular spur gears comprising triptycene rotators and bibenzimidazole stators. *J. Am. Chem. Soc.* **2012**, *134*, 1528–1535.

97. H. Ube, R. Yamada, J. Ishida, H. Sato, M. Shiro and M. Shionoya, A circularly arranged sextuple triptycene gear molecule. *J. Am. Chem. Soc.* **2017**, *139*,16470–16473.

98. G. Yamamoto, and M. Oki, Correlated rotation in 9-(2,4,6-trimethylbenzyl)-triptycenes. Direct and roundabout enantiomerization and diastereomerization processes. *J. Org. Chem.* **1983**, *48*, 1233–1236.

99. R.B. Nachbar Jr., W.D. Hounshell, V.A. Naman, O. Wennerström, A. Guenzi and K. Mislow, Application of empirical force field calculations to internal dynamics in 9-benzyltriptycenes. *J. Org. Chem.* **1983**, *48*, 1227–1232.

100. We note, however, that in 9-(ferrocenylmethyl)triptycene the bulky sandwich unit cannot fit into the valley between the blades resulting in directional rotational preference for the two enantiomers *when locally traversing a single blade.* K. Nikitin, Y. Ortin and M.J. McGlinchey, Dynamics of a molecular rotor exhibiting local directional preference within each enantiomer. *J. Phys. Chem.* **2021**, *125*, 2061–2068.

101. K. Nikitin, H. Müller-Bunz, Y. Ortin, J. Muldoon and M.J. McGlinchey, Molecular dials: hindered rotations in mono- and di-ferrocenyl anthracenes and triptycenes. *J. Am. Chem. Soc.* **2010**, *132*, 17617–17622.

102. K. Nikitin, J. Muldoon, H. Müller-Bunz and M.J. McGlinchey, A ferrocenyl kaleidoscope: slow interconversion of six diasteroatropisomers of 2,6-di-*tert*-butyl-9,10-diferrocenyltriptycene. *Chem. Eur. J.* **2015**, *21*, 46640–4670.

103. T.R. Kelly, Progress toward a rationally designed molecular motor. *Acc. Chem. Res.* **2001**, *34*, 514–522.

104. A.M. Stephen and C.J. Richards, A metallocene molecular gear. *Tetrahedron Lett.* **1997**, *38*, 7805–7808.

105. L.E. Harrington, L. Cahill and M.J. McGlinchey, Towards an organometallic molecular brake with a metal foot-pedal: the synthesis, dynamic behavior and X-ray crystal structure of (9-indenyl)triptycene-Cr(CO)$_3$. *Organometallics* **2004**, *23*, 2884–2891.

106. Y.F. Oprunenko, Inter-ring haptotropic rearrangements in pi complexes of transition metals with polycyclic aromatic ligands. *Russ. Chem. Rev.* **2000**, *69*, 683–704.

107. K. Nikitin, H. Müller-Bunz, Y. Ortin and M.J. McGlinchey, A molecular pad-dle-wheel with a sliding organometallic latch: syntheses, X-Ray crystal struc-tures and dynamic behaviour of [η⁶-2-(9-triptycyl)-indene]Cr(CO)₃, and of [η⁵-2-(9-triptycyl)-indenyl]M(CO)₃, M = Mn, Re. *Chem. Eur. J.* **2009**, *15*, 1836–1843.
108. K. Mislow, Stereochemical consequences of correlated rotation in molecular propellers. *Acc. Chem. Res.* **1976**, *9*, 26–33.
109. H.B. Bürgi and J.D. Dunitz, From crystal statics to chemical dynamics. *Acc. Chem. Res.* **1983**, *16*, 153–161.
110. H.B. Bürgi, J.D. Dunitz, J.M. Lehn and G. Wipff, Stereochemistry of reaction paths at carbonyl centres. *Tetrahedron* **1974**, *30*, 1563–1572.
111. R.H. Crabtree and M. Lavin, Structural analysis of the semibridging carbonyl. *Inorg. Chem.* **1986**, *25*, 805–812.
112. L. Girard, P.E. Lock, H. El. Amouri and M.J. McGlinchey, The rearrangement pathway in $[Cp_2Mo_2(CO)_4(RC{\equiv}C\text{-}CR_2)]^+$ cations: an extended Hückel molec-ular orbital and Bürgi-Dunitz trajectory study. *J. Organomet. Chem.* **1994**, *478*, 189–196.
113. E. Bye, W.D. Schweizer and J.D. Dunitz, Chemical reaction paths. 8. Stereoisomerization path for triphenylphosphine oxide and related molecules: indirect observation of the structure of the transition state. *J. Am. Chem. Soc.* **1982**, *104*, 5893–5898.
114. J.M. Chance, J.H. Geiger and K. Mislow, A parity restriction on dynamic gearing immobilizes the rotors in tris(9-triptycyl)germanium chloride and tris(9-triptycyl)cyclopropenium perchlorate. *J. Am. Chem. Soc.* **1989**, *111*, 2326–2327.

Chapter 6

A Miscellany of Periodic Table Relationships

"We build too many walls and not enough bridges."

— *Isaac Newton*

This short chapter is intended to provide a conceptual bridge between the major themes of breaking and making molecular symmetry. In the first section, we describe the syntheses (or attempts to prepare) compounds in which their symmetry is systematically lowered by the sequential replacement of atoms or substituents by different elements, positioned directly adjacent either across or down the Periodic Table. Multiple bonds between metal and ligands or directly between metals are also discussed. In the second section, we indicate how the apparent randomness of NMR spin-spin coupling constants (J values) as we progress across or down the Periodic Table hides an underlying regularity when expressed as reduced coupling constants (K values).

6.1. Chiral Methanes

The quest for simple chiral molecules, such as polyhalogenated methanes, possessing only single bonds and lacking conformational mobility has been the focus of widespread interest for more than 120 years.

The great pioneer was Frédéric Swarts (1866–1940), who worked at the University of Ghent in Belgium and eventually became a professor as one of the successors to Kekulé of benzene ring fame. He developed convenient routes to organofluorine compounds by using inorganic reagents such as bromine and antimony trifluoride (Swarts' reaction), and successfully prepared many chlorofluoro- and bromofluoromethanes, including bromochlorofluoromethane in 1893.[1]

This project was taken up in 1942 by Berry and Sturtevant at Yale University, whose aim was to resolve bromochlorofluoromethane into its enantiomers. Their approach was to brominate chloroacetal which, upon hydrolysis, yielded chlorodibromomethane; subsequent treatment with bromine and SbF_3 furnished bromochlorofluoromethane which they obtained in 140 g yield and carefully fractionated to obtain 37 g of pure material. Attempts to separate the enantiomers by formation of addition compounds with chiral substances such as digitonin were, however, unsuccessful.[2]

In 1952, a different approach was taken by Robert Haszeldine, then a young researcher at Cambridge, subsequently a professor in Manchester and a major contributor to many aspects of fluorine chemistry, who used the Hunsdiecker reaction to generate mixed polyhalogenated methanes.[3] Typically, the reaction of silver dichlorofluoroacetate, $CFCl_2CO_2Ag$, with I_2 gave $CFCl_2I$ and AgI; this method allowed the ready preparation of many other such methanes, e.g., $CFClBr_2$, $CFCl_2Br$, and CHFBrI. However, the relative weakness of the carbon-iodine bond precluded the isolation of bromochlorofluoroiodomethane, CFClBrI, a situation that still exists today. Bromochlorodifluoromethane, CF_2ClBr, later known as Halon 1211 or Freon 12B1, is widely used as a fire suppressant, especially for expensive electrical equipment. Other Freons were adopted worldwide as refrigerants until their effect on the ozone layer was established, and their use has been largely discontinued.

The first, at least partially, successful separation of the enantiomers of CHFClBr was achieved in 1969 by Hargreaves and Modarai by hydrolysis of 1,1,1-bromochlorofluoroacetone, $FClBrC–C(=O)CH_3$, which itself had been resolved by fractional crystallisation of its (-)-menthyl-*N*-aminocarbamate derivative.[4] The reported rather low

$[\alpha]^{19}_{D}$ value of 0.25°, recorded in cyclohexane, was later shown to be the result of incomplete separation of the enantiomers.

Some years later, this approach was modfied by Wilen *et al.* using the formation of inclusion compounds with the chiral alkaloid brucine;[5] however, the enantiomeric excess (*ee*) was only 4.3% as established by an NMR method, which had just then (1985) been developed, and that used enantioselective inclusion into a tailor-made cryptophane.[6] Although the observed $[\alpha]^{25}_{D}$ value of 0.128° was clearly lower that that achieved by Hargreaves and Modarai, since the *ee* could be determined relatively accurately this corresponded to an $[\alpha]^{25}_{D}$ value of ~3.0° for the pure enantiomer. In 1989 the synthesis of CHFClBr was considerably improved by fractional crystallisation of the strychnine salt of bromochlorofluoroacetic acid to yield (+)-CHFBrCl with an optical purity of 66%. This gave an $[\alpha]^{25}_{D}$ value for the pure enantiomer of 2.75 ± 0.05°.[7] Almost 20 years later, the exact same sample originally obtained by Wilen (in New York) was re-examined by Collet (in Paris) who used capillary gas chromatography with a chiral stationary phase to separate the enantiomers; gratifyingly, the *ee* value of 4.3 ± 1% was unambiguously confirmed.[8]

At this point, the project became a significant focus in the physics community. Based on the assumption that the polarisabilities decrease in the order Br > Cl > H > F, it had been postulated that the (*S*) antipode should have positive rotations for a range of wavelengths. Interestingly, according to the results of high-level density functional theory (DFT) calculations, the absolute configurations of CHFClBr are indeed (*S*)-(+) and (*R*)-(-), as illustrated in Figure 6.1.[9]

In another important development, it has been suggested that a parity violation effect induced by the nuclear weak interaction could

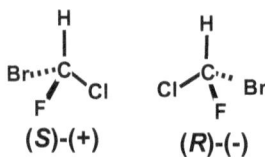

Figure 6.1. Enantiomers of bromochlorofluoromethane.

lead to a minuscule energy difference in the vibrational spectra of the enantiomers of CHFClBr. Thus, according to this hypothesis (which is still somewhat controversial), these enantiomers would cease to be exact mirror images of each other. Preliminary experiments have not yet reached the level of precision required, but the concept provides an exciting link between synthetic chemistry and subnuclear physics.[10]

Finally, in an extraordinary experimental breakthrough, it has been shown that the absolute molecular stereochemistry of CHFClBr can be determined directly by gas phase coulomb explosion imaging. In this novel approach, a supersonic gas jet of the chiral molecule crosses a high-power femtosecond laser that induces multiple ionisation. The highly charged ion suffers coulombic explosion and the masses and velocities of the resulting fragments are analysed on a position- and time-sensitive multichannel plate detector. The mass assignment, and therefore the identity of the isotopomer, be it $CHF^{35}Cl^{79}Br$, $CHF^{35}Cl^{81}Br$, $CHF^{37}Cl^{79}Br$, or $CHF^{37}Cl^{81}Br$, is readily apparent. Enantiomers are distinguished by defining an angle θ, and then indicating whether the momenta of the bromine, chlorine and fluorine form a right- or left-handed coordinate system. This technique has also been used to distinguish the isotopically chiral enantiomers of $HCBr^{35}Cl^{37}Cl$.[11]

6.2. Chiral Mixed-metal Clusters

Focussing initially on transition metal carbonyls, we note that while homo-dinuclear complexes of the type $M_2(CO)_{10}$, where M = manganese, technetium or rhenium, $M_2(CO)_9$, where M = iron, ruthenium or osmium, and $M_2(CO)_8$, where M = cobalt (and rhodium or iridium, at least under matrix isolation conditions), are regularly encountered in the literature, analogous hetero-dinuclear systems such as $MnCo(CO)_9$ are much less common. Indeed, molecules of the type $MM'(CO)_n$, where n is an even number and M and M' are in consecutive Periodic Table groups, would contravene the Effective Atomic Number rule whereby pairs of 17-electron molecular fragments can couple to form stable compounds. However, mixed-metal trinuclear complexes have been systematically prepared in which the Periodically adjacent metals form a chain. Typically, photolysis of manganese and iron carbonyls yields $(OC)_5Mn–Fe(CO)_4–Mn(CO)_5$ and chlorination of cyclic $Ru_2Os(CO)_{12}$ leads to $ClRu(CO)_4–Os(CO)_4–Ru(CO)_4Cl$; both have been fully characterised by X-ray crystallography.[12,13]

Many of the most impressive achievements in the field of mixed metal clusters came from the laboratory of Heinrich Vahrenkamp at the University of Freiburg, near the Black Forest in the south of Germany. He and his co-workers developed logical routes to many such complexes, in particular those containing Periodic Table neighbours, as well as preparing, separating and fully characterising the enantiomers of chiral clusters.[14]

In a typical process, depicted in Scheme 6.1, the reaction of $MePCl_2$ with $Fe_2(CO)_9$ yields the complex, **1**, in which the phosphine is now coordinated to the $Fe(CO)_4$ fragment. Nucleophilic displacement of a halide by the tetracarbonylcobalt anion forms the phosphorus-iron-cobalt ring, **2**. This is followed by a second nucleophilic displacement using the pentacarbonylmanganese anion, thus generating the trimetallic compound, **3**, that readily loses two carbon monoxide ligands to give the tetrahedral cluster, **4**, in which the first row transition metals from Groups 7, 8 and 9 are linked.[15] It is also possible to introduce chromium from Group 6 as a third metal by using the $(C_5H_5)Cr(CO)_3$ anion as the attacking nucleophile. Moreover, the closely analogous sulfur-capped mixed-metal tetrahedral clusters containing elements such as Cr, Mn, Fe, Co, Mo and W have also been prepared.[16]

A second synthetic approach involved direct replacement of a $Co(CO)_3$ vertex in a "Seyferth-type" carbynyltricobalt cluster,

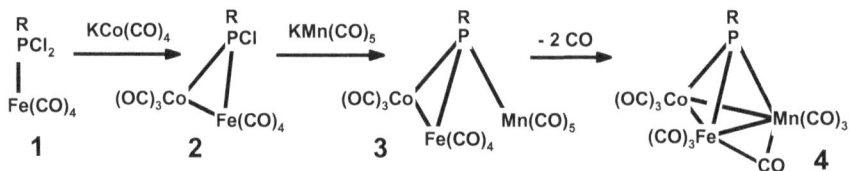

Scheme 6.1.   Generation of phosphorus-capped Mn-Fe-Co tetrahedral clusters.

$RCCo_3(CO)_9$, by an isolobal moiety[17] such as $Mn(CO)_4$ or (C_5H_5) Ni to produce chiral tetrahedral clusters. Surprisingly, perhaps, the reaction of $Co_3(CO)_9C-CO_2Me$ with $[(C_5H_5)Ni(CO)]_2$ brings about not merely a second substitution of cobalt by nickel to form the $CoNi_2$ cluster, **5**, but, in low yield, complete replacement to form $(C_5H_5)_3Ni_3C-CO_2Me$, **6** (Scheme 6.2).[18] This direct displacement technique has also been applied to $(alkyne)Co_2(CO)_6$ clusters with subsequent addition of an $Fe(CO)_3$ cap whereby the tetrahedron has expanded to form a square-based pyramidal complex, **7**, containing adjacent metals from Groups 8, 9 and 10 (Scheme 6.3).[19]

Turning now to the question of the resolution of the enantiomers of these trimetallic tetrahedral clusters, once again this was first accomplished by Vahrenkamp, as indicated in Scheme 6.4. Treatment of the racemic sulfur-capped cluster, **8**, with an optically pure phosphine was followed by separation of the resulting diastereomers (of optical purity ~98%) by crystallisation; subsequent removal of the phosphine yielded enantiomers of **8**. The optical rotations of these chiral clusters are extreme such that the molar rotations Φ are in the range 10,000° to 40,000°.[20] As noted in Chapter 5, the absolute

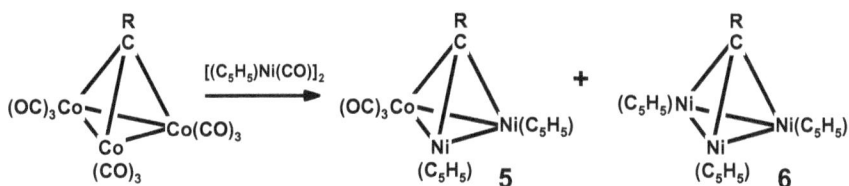

Scheme 6.2. Formation of carbyne-capped trimetallic tetrahedral clusters.

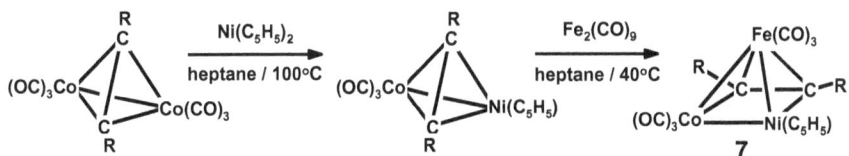

Scheme 6.3. Formation of Fe-Co-Ni square-based pyramidal clusters.

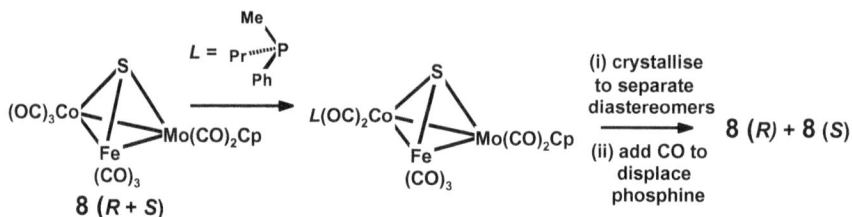

Scheme 6.4. Separation of the enantiomers of sulfur-capped trimetallic tetrahedral clusters.

configuration of such molecules can be assigned by designating an *R* or *S* label to a dummy atom placed inside the tetrahedron, and then following the usual Cahn-Ingold-Prelog rules. Thus, the structures shown in Scheme 6.4 have the *R* configuration (Mo > Co > Fe > S).

6.3. Sequential Multiple Bonds to a Single Metal Centre

In a variant of this approach, there are a number of molecules in which a metal is bonded to the same element, but with sequentially increasing bond order. Typically, Schrock reported the synthesis of a tungsten complex, **9**, bearing alkyl, alkylidene and alkylidyne ligands.[21] The structure was determined by Churchill and revealed that the tungsten-carbon distances for the single, double and triple linkages were 2.258(8), 1.942(9) and 1.785(8) Å, respectively.[22] In a somewhat analogous anionic chromium complex $[Cr\equiv N(=NPh)(N^iPr_2)_2]^-$, **10**, exhibiting metal-nitrogen single, double and triple bonds, the pattern of gradually decreasing bond lengths is repeated: Cr–N 1.879(3), Cr=N 1.728(3), and Cr≡N 1.554(3) Å.[23]

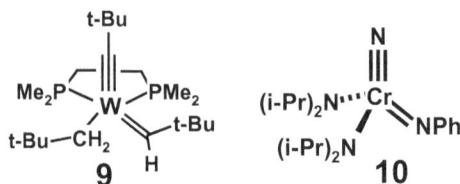

6.4. Metal-metal Multiple Bonds

Extending these concepts to interactions directly between metal centres, we note that single bonds between transition metals in carbonyl complexes such as $(OC)_5M–M(CO)_5$, where M = Mn or Re, have been known for at least eight decades,[24] and were first structurally characterised in 1957.[25] However, metal-metal multiply-bonded systems are of more recent vintage, especially in cases of enhanced bond order.

It was already known that the reaction of cyclopentadiene and molybdenum hexacarbonyl led to $(C_5H_5)(OC)_3Mo–Mo(CO)_3(C_5H_5)$ in which the 18-electron rule is satisfied. However, the corresponding reaction using pentamethylcyclopentadiene (Cp*H) led instead to a molecule with the formula $(C_5Me_5)_2(OC)_4Mo_2$, which led Bruce King to suggest the presence of a molybdenum-molybdenum triple bond, as in **11**.[26] This proposal was subsequently confirmed by an X-ray crystallographic study on the analogous chromium complex $Cp*_2(CO)_4Cr_2$, which revealed a very short Cr–Cr distance of 2.276 Å, strongly indicating the existence of a chromium-chromium triple bond.[27] Since that time, an enormous number of inorganic and organometallic complexes containing metal-metal triple bonds have been reported. In particular, the pioneering work in this field was dominated by F. Albert Cotton and his many collaborators, initially at MIT and later at Texas A&M.[28] Metal-metal double bonds have also been widely studied. A typical early instance is the rhodium dimer $Cp*Rh(\mu\text{-}CO)_2RhCp*$, **12**,[29] while a much more recent example is shown in Figure 6.2, whereby a $ZnCl_2$ unit bridges a ruthenium-ruthenium double bond in **13**.[30]

The important feature of these multiple bonds is the pivotal role played by orbital overlap between *d* orbitals. Thus, interactions

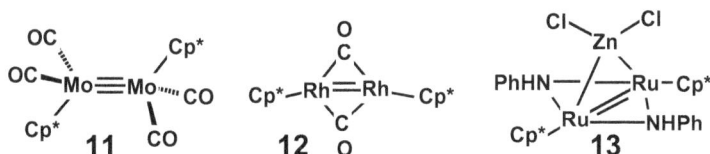

Figure 6.2.   Selected molecules possessing metal-metal triple or double bonds.

between d_{z^2} orbitals along the internuclear axis give rise to σ bonds, whereas π bonds arise from overlap involving d_{xy}/d_{xy} and d_{xz}/d_{xz} combinations (Figure 6.3); as usual, the out-of-phase combinations define the metal-metal σ* and π* molecular orbitals. In this fashion, triple bonds are formed by six electrons in a $\sigma^2\pi^4$ ground state, entirely analogous to the bonding in alkynes, except using d orbitals rather than s and p combinations.

However, a major breakthrough was the report by Cotton that the anion in $K_2[Re_2Cl_8]$, **14**, adopted D_{4h} geometry and exhibited a remarkably short rhenium-rhenium distance of 2.24 Å.[31] This was categorised as the first example of a quadruple metal-metal linkage, and invoked the concept of a δ bond whereby the d_{xy} orbital on each metal interacts with its counterpart in a face-to-face manner, as depicted in Figure 6.3. Since such a geometric arrangement necessarily leads to smaller overlap integrals, the splitting between the δ and δ* energy levels is rather small and indeed results in absorption in the visible region. Thus, the $[Re_2Cl_8]^{2-}$ and isoelectronic $[Mo_2Cl_8]^{4-}$ salts exhibit δ→δ* transitions from their singlet $\sigma^2\pi^4\delta^2$ ground states to give a singlet $\sigma^2\pi^4\delta\delta^*$ excited state, and are royal blue and intensely red, respectively. The rhenium complex, **14**, was also independently prepared by Russian workers and the structure was featured on a postage stamp commemorating a chemical congress.

Since there are five d orbitals, one might venture to suggest that the quintuple metal-metal bond should be possible. As illustrated in Figure 6.4, the d_{xy} and $d_{x^2-y^2}$ orbitals, which are oriented at 45° to each

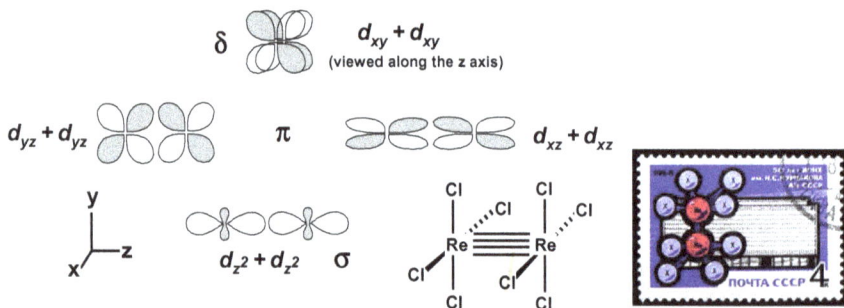

Figure 6.3.   Orbital interactions in metal-metal triple and quadruple bonds.

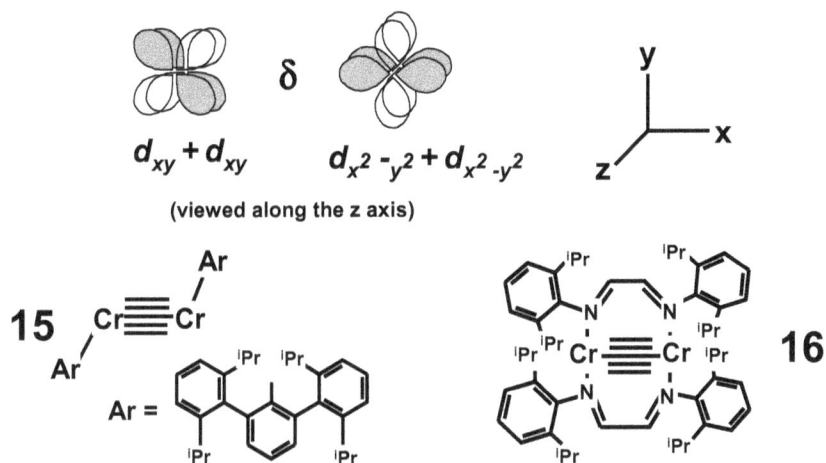

Figure 6.4.   Molecules possessing metal-metal quintuple bonds.

other, are in principle both capable of forming δ bonds. However, in the D_{4h}-symmetric quadruply bonded systems, the $d_{x^2-y^2}$ orbital is normally heavily involved in metal-ligand bonding and so is not available to participate in a second δ interaction. Nevertheless, this concept first reached experimental fruition in the laboratory of Philip Power at the University of California, Davis. In the *trans-bent* diarene-dichromium molecule, **15**, where the arene ligand is exceptionally bulky, the separation between the two d^5 metal centres was found to be only 1.8351(4) Å, and this remarkably short bond, together with the observed magnetic data and high-level calculations, strongly indicated the existence of five-fold metal-metal bonding.[32] Since that time, a number of other molecules possessing quintuple metal-metal bonds have been characterised; typically, in the diazadiene complex, **16**, whereby the ligands bridge the two metals, the Cr–Cr distance is extremely short at 1.8028(9) Å.[33] This area has been comprehensively reviewed.[34,35] For a number of years, there has also been much discussion about the assignment of bond order in such species as Mo_2, W_2 or U_2; arguments have been made for the existence of the sextuple bond, but there is still disagreement among the community of theoreticians.[36]

6.5. A Pseudo-butane made up of Consecutive Group 14 Elements

The Group 14 elements, carbon, silicon, germanium and, to a lesser extent, tin and lead, are distinguished by their ability to catenate, i.e., form chains. The hydrocarbons, of course, are numbered in the millions, but polysilanes or germanes, such as Si_6H_{14} and Ge_9H_{20}, are known, and these elements also form many rings and long-chain polymers. The chemistry of these species, in particular those involving linkages between the different group members, has for many years been extensively explored by Keith Pannell at the University of Texas at El Paso. Among his achievements is the synthesis of the remarkable tetra-substituted methane $C(SiMe_3)(GeMe_3)(SnMe_3)(PbMe_3)$, **17**; however, what might be considered the *pièce de résistance* is his achievement in linking the first four of these elements in Periodic sequence.

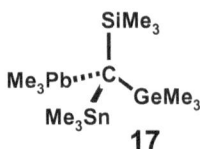

$$SiMe_3$$
$$|$$
$$Me_3Pb\cdots C$$
$$\diagdown GeMe_3$$
$$\diagup$$
$$Me_3Sn$$
$$\mathbf{17}$$

Starting with *tert*-butyllithium, this was used to displace a chloride from dichlorodimethysilane to form $Me_3C–SiMe_2Cl$, **18**. Subsequent reaction of **18** with $PhMe_2GeLi$ furnished $Me_3C–SiMe_2–GeMe_2Ph$, **19** (Scheme 6.5). The incorporation of the phenyl substituent in the germanium reagent was important since it is relatively easily displaced to form the Ge–Cl bond in

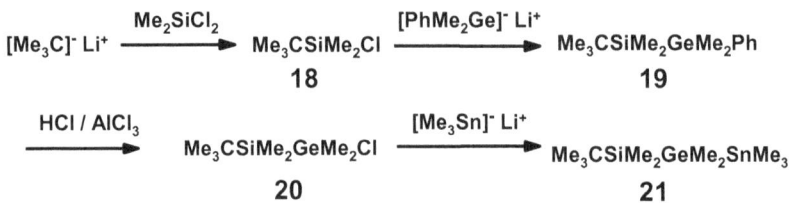

$$[Me_3C]^- Li^+ \xrightarrow{\ Me_2SiCl_2\ } Me_3CSiMe_2Cl \xrightarrow{\ [PhMe_2Ge]^- Li^+\ } Me_3CSiMe_2GeMe_2Ph$$
$$\mathbf{18} \qquad\qquad \mathbf{19}$$

$$\xrightarrow{\ HCl\ /\ AlCl_3\ } Me_3CSiMe_2GeMe_2Cl \xrightarrow{\ [Me_3Sn]^- Li^+\ } Me_3CSiMe_2GeMe_2SnMe_3$$
$$\mathbf{20} \qquad\qquad \mathbf{21}$$

Scheme 6.5.   Synthetic route to $Me_3C–SiMe_2–GeMe_2–SnMe_3$, **21**.

$Me_3C–SiMe_2–GeMe_2Cl$, **20**, required for the final step. To complete the synthesis, **20** was treated with R_3SnLi, where R = methyl or phenyl, and the desired product $Me_3C–SiMe_2–GeMe_2–SnR_3$, **21**, was obtained and fully characterised spectroscopically. The X-ray structure revealed a gradual increase in bond length between successively heavier elements: carbon-silicon 1.900, silicon-germanium 2.386, and germanium-tin 2.608 Å.[37]

6.6. Reduced Spin-spin Coupling Constants in NMR

Having discussed examples of the systematic formation of linkages between adjacent elements, we now turn our attention to the apparently random distribution of NMR spin-spin coupling constants as we traverse the Periodic Table. In fact, anyone who has run a ^{13}C NMR spectrum has already encountered this phenomenon. Such spectra are commonly acquired in deuterochloroform, $CDCl_3$, as the solvent; it gives rise to an intense 1:1:1 triplet attributable to coupling between carbon-13 and deuterium (which has a spin quantum number $I = 1$). Although one normally ignores the solvent signal, and takes little notice of the value of this coupling constant for which $^1J_{C–D} = 32$ Hz, it should be compared to the corresponding value $^1J_{C–H} = 209$ Hz in $CHCl_3$. Likewise, the solvent resonance for $^{13}CD_2Cl_2$ appears as a 1:2:3:2:1 quintet with $^1J_{C–D} = 27.3$ Hz, whereas in CH_2Cl_2, $^1J_{C–H} = 178$ Hz.

These differences arise as the result of their inherent nuclear properties whereby the magnetogyric factors, γ_H and γ_D, differ by a factor of 6.514. This factor is, of course, also reflected in the ratio of spectrometer frequencies at which these nuclei are acquired; on a 100 MHz spectrometer whereby the protons in tetramethylsilane resonate at 100 MHz, the corresponding value for its deuterium analogue, $(CD_3)_4Si$, is 15.351 MHz.

Since these interactions between nuclei X and Y are transmitted primarily by the electrons, to compare coupling in analogous molecules one must remove the nuclear dependence of $^1J_{XY}$ by introducing the reduced coupling constant $^1K_{XY}$, defined as:

$^nK_{XY} = {}^nJ_{XY} (4\pi^2/h\, \gamma_X\, \gamma_Y)$ where n is the number of bonds between X and Y.

For $^nJ_{XY}$ measured in Hz, the units of $^nK_{XY}$ in SI units are $NA^{-2}m^{-3}$ (newtons ampere^{-2} m^{-3}), and usually possess values in the range 10^{20}–10^{22} $NA^{-2}m^{-3}$.

Although very well known to NMR spectroscopists, reduced coupling constants are less well appreciated in the general literature and can be a source of error when comparing the magnitudes of experimentally observed J_{XY} values between different pairs of nuclei. Typically, as depicted in Figure 6.5, the observed $^1J(^{45}Sc–^{19}F)$ value of 167 Hz in the $[ScF_6]^{3-}$ anion would appear to be considerably larger than the observed $^1J(^{183}W–^{19}F)$ value of 44.1 Hz in WF_6. However, the reduced coupling constants are actually reversed: $^1K_{W–F}$ is 9.39 × 10^{20} while $^1K_{Sc–F}$ is only 6.09 × 10^{20} $NA^{-2}m^{-3}$; it is now apparent that

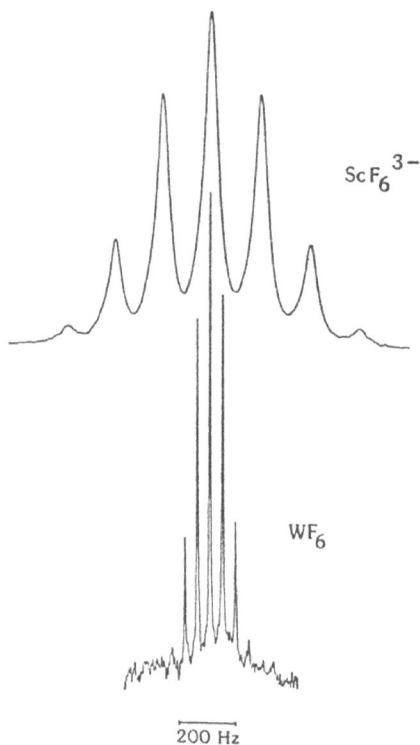

Figure 6.5. ^{45}Sc (upper) and ^{183}W (lower) NMR spectra exhibiting 1:6:15:20:15:6:1 septets arising from spin-spin coupling to six equivalent ^{19}F nuclei. (Data taken from the PhD thesis of D.G. Bickley, McMaster University, 1982.)

the tungsten-fluorine coupling is actually 54% larger than the scandium-fluorine coupling in these analogous octahedral hexafluoro derivatives.

In fact, a quick and easy way to determine the relative magnitudes of the spin-spin couplings in such structurally similar systems, where the identity of only a single nucleus has changed, is simply to divide each J_{XY} value by the spectrometer frequency used for its acquisition. Thus, if the WF_6 and $[ScF_6]^{3-}$ spectra were both to be recorded on a 100 MHz instrument, the resonance frequencies for ^{183}W and ^{45}Sc and would be 4.17 MHz and 24.294 MHz, respectively. Now, using the $^1J(^{183}W-^{19}F)$ and $^1J(^{45}Sc-^{19}F)$ values 44.1 Hz and 167 Hz listed above, $44.1/4.17 = 10.57$ and $167/24.294 = 6.874$; the ratio 10.57:6.84 reveals the tungsten-fluorine coupling to be 54% larger.

Another illustrative example is provided by comparing the nitrogen-hydrogen couplings in various isotopomers of the ammonium ion, NH_4^+. The values of $^1J(^{15}N-^1H)$, $^1J(^{14}N-^1H)$ and $^1J(^{15}N-D)$ are 73.3, 52.23 and 8.02 Hz, respectively, from which no clear pattern is evident. However, the corresponding reduced coupling constants $^1K(^{15}N-^1H)$, $^1K(^{14}N-^1H)$ and $^1K(^{15}N-D)$ are 6.029×10^{20}, 6.025×10^{20} and 6.027×10^{20} $NA^{-2}m^{-3}$, respectively; they are identical within experimental error!

6.7. Making K_{AB} Patterns out of Apparent J_{AB} Chaos

In light of the above observations, whereby the effect of the nuclear magnetogyric factors have been removed, one can now search for systematic correlations as we move sequentially either horizontally or vertically within the Periodic Table. The $^1J_{M-H}$ values for the group 14 hydrides CH_4, SiH_4, GeH_4 and SnH_4 are 125, 202.5, 97.6 and 1930 Hz, respectively. Although plumbane, PbH_4, is too unstable to yield a reliable coupling constant, the values of $^1J_{Pb-H}$ in $(CH_3)_2PbH_2$ and $(CH_3)_3PbH$ were found to be 2455 and 2380 Hz, respectively. There appears to be no pattern here; however, the corresponding $^1K_{M-H}$ values for the C, Si, Ge and Sn hydrides are 4.14×10^{20}, 8.49×10^{20}, 23.3×10^{20}, and 43.1×10^{20} $NA^{-2}m^{-3}$.

As was pointed out by Len Reeves from the University of British Columbia, Vancouver, as long ago as 1964,[38] there is an excellent

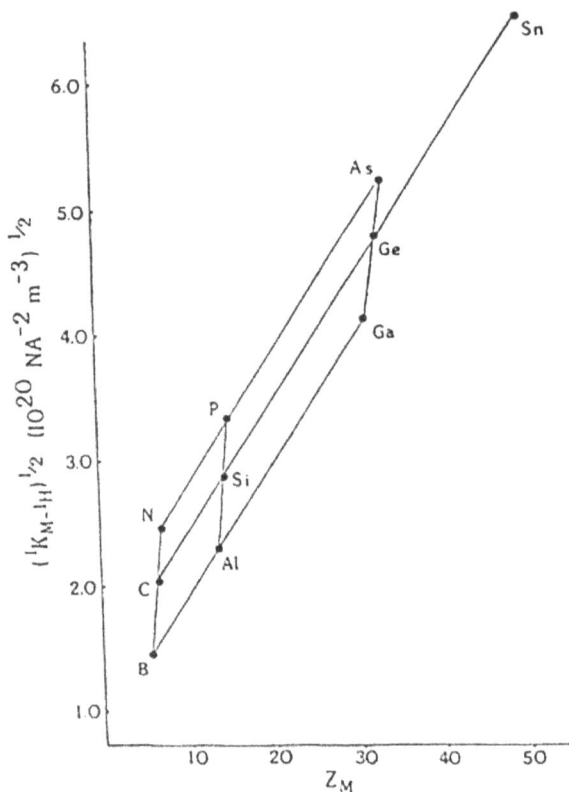

Figure 6.6. Correlation of the reduced coupling constants, $(^1K_{M-H})^{1/2}$, for the ions or molecules $[MH_4]^n$ ($n = -1, 0$ or $+1$) in Groups 3, 14 and 15 as a function of atomic number Z_M.

correlation of the atomic number of the central element, Z_M, with the square root of the reduced coupling constant, $[^1K(M-^1H)]^{1/2}$. This was interpreted in terms of the dominance of the Fermi contact term, which is dependent on the electron density at the nucleus. Similar correlations are found for the series of tetrahedral anions $[BH_4]^-$, $[AlH_4]^-$, and $[GaH_4]^-$, and also for the corresponding cations $[NH_4]^+$, $[PH_4]^+$, and $[AsH_4]^+$. Figure 6.6 shows a parallelogram plot incorporating all these data.

Since that time, many such correlations have been reported. These include the $^1K_{M-C}$ values in the series $M(CH_3)_4$, $M = C$, Si, Ge, Sn and Pb, with a correlation coefficient of 0.999, as well as numerous examples involving Periodic series of main group and transition metal hydrides, fluorides, oxides and other elements. However, our goal here is not to present a comprehensive listing of such data, but rather to note that merely comparing J values and drawing conclusions about the relative strength of bonding interactions can be contentious. Gratifyingly, the most recent calculations of reduced spin-spin coupling constants very successfully match experimental results, thus giving us great confidence in their predictions for currently unknown molecules.[39]

References

1. G.B. Kauffman, Frédéric Swarts: pioneer in inorganic fluorine chemistry. *J. Chem. Ed.* **1955**, *32*, 301–303.
2. K.L. Berry and J.M. Sturtevant, Fluorochlorobromomethane. *J. Am. Chem. Soc.* **1942**, *64*, 1599–1600.
3. R.N. Haszeldine, The reactions of metallic salts of acids with halogens. Part III. Some reactions of salts of fluorohalogenoacetates and of perfluoroacids. *J. Chem. Soc.* **1952**, 4259–4268.
4. M.K. Hargreaves and B. Modarai, An optically active haloform: (+)-bromochlorofluoromethane. *Chem. Commun.* **1969**, 16.
5. S.H. Wilen, K.A. Bunding, C.M. Kascheres and M.J. Wieder, On the optical activity of bromochlorofluoromethane. *J. Am. Chem. Soc.* **1985**, *107*, 6997–6998.
6. J. Canceill, L. Lacombe and A. Collet, Analytical optical resolution of bromochlorofluoromethane by enantioselective inclusion into a tailor-made "cryptophane" and determination of its maximum rotation. *J. Am. Chem. Soc.* **1985**, *107*, 6993–6996.
7. T.R. Doyle and O. Vogl, Bromochlorofluoromethane and deuteriobromochlorofluoromethane of high optical purity. *J. Am. Chem. Soc.* **1989**, *111*, 8510–8511.
8. H. Grosenick, V. Schurig, J. Costante and A. Collet, Gas chromatographic enantiomer separation of bromochlorofluoromethane. *Tetrahedron Asymmetry* **1995**, *6*, 87–88.
9. P.L. Polavarapu, The absolute configuration of bromochlorofluoromethane. *Angew. Chem. Int. Ed.* **2002**, *41*, 4544–4546.
10. J. Crassous, F. Monier, J-P. Dutasta, M. Ziskind, C. Daussy, C. Grain and C. Chardonnet, Search of resolution of chiral fluorohalogenomethanes and parity-violation effects at the molecular level. *ChemPhysChem* **2003**, *4*, 541–548.

11. M. Pitzer, M. Kunitski, A.S. Johnson, T. Jahnke, H. Sann, F. Sturm, L.P.H. Schmidt, H. Schmidt-Böcking, R. Dörner, J. Stohner, J. Kiedrowski, M. Reggelin, S. Marquardt, A. Schiesser, R. Berger and M. Schöffler, Direct determination of absolute molecular stereochemistry in gas phase by coulomb explosion imaging. *Science* **2013**, *341*, 1096–1099.

12. P. Agron, R. Ellison and H.A. Levy, The crystal structure of dimanganese iron carbonyl, $Mn_2Fe(CO)_{14}$. *Acta Crystallogr.* **1967**, *23*, 1079–1086.

13. P. Hirva, M. Haukka, M. Jakonen and T.A. Pakkanen, Growth of the metal framework in linear ruthenium and osmium carbonyls. *Inorg. Chim. Acta* **2006**, *359*, 853–862.

14. H. Vahrenkamp, Framework chirality an optical activity of organometallic cluster compounds. *J. Organomet. Chem.* **1989**, *370*, 65–73.

15. M. Müller and H. Vahrenkamp, Cluster construction: stepwise build-up of μ_3-RP-trimetallic clusters via P-halogen compounds. *Chem. Ber.* **1983**, *116*, 12322–2336.

16. F. Richter and H. Vahrenkamp, Chiral SFeCoM clusters: preparation, reactivity and establishment of chirality. *Chem. Ber.* **1982**, *115*, 3224–3242.

17. Two fragments are isolobal with each other if the number, symmetry properties, approximate energy and shape of the frontier orbitals, and number of electrons in them are similar. R. Hoffmann, Building bridges between inorganic and organic chemistry (Nobel Lecture). *Angew. Chem. Int. Ed.* **1982**, *21*, 711–724.

18. R. Blumhofer, K. Fischer and H. Vahrenkamp, Multiple metal exchange in μ_3-methylidyne tricobalt clusters. *Chem. Ber.* **1986**, *119*, 104–214.

19. M. Mlekuz, P. Bougeard, B.G. Sayer, M.J. McGlinchey, S. Peng, A. Marinetti, J-Y. Saillard, J. Ben Naceur, B. Mentzen and G. Jaouen, Syntheses, crystal structures and DNMR studies on the mixed clusters $CpNiFe(CO)_3(RC°CR')M$, (M = CpNi, $Co(CO)_3$, $Mo(CO)_2Cp$): some comments on the acetylene rotation process. *Organometallics* **1985**, *4*, 1123-1130.

20. F. Richter and H. Vahrenkamp, The first optically active cluster: resolution of enantiomers and absolute configuration of $SFeCoMoCp(CO)_8$. *Angew. Chem. Int. Ed. Engl.* **1980**, *19*, 65.

21. D.N. Clark and R.R. Schrock, Multiple metal-carbon bonds. 12. Tungsten and molybdenum neopentylidyne and some tungsten neopentylidene complexes. *J. Am. Chem. Soc.* **1978**, *100*, 6774–6776.

22. M.R. Churchill and W.J. Youngs. Crystal structure and molecular geometry of $W(\equiv CCMe_3)(=CHMe_3)(CH_2CMe_3)(dmpe)$, a molecular tungsten(VI) complex with metal-alkylidyne, metal-alkylidene, and metal-alkyl linkages. *Inorg. Chem.* **1979**, *18*, 2454–2458.

23. E.P. Beaumier, B.S. Billow, A.K. Singh, S.M. Biros and A.L. Odom, A complex with nitrogen single, double and triple bonds to the same chromium atom: synthesis, structure and reactivity. *Chem. Sci.* **2016**, *7*, 2532–2536.

24. W. Hieber and H. Fuchs, Metal carbonyls XXXVIII. Rhenium pentacarbonyl. *Z. Anorg. Allgem. Chem.* **1941**, *248*, 256–268.

25. L.F. Dahl, E. Ishishi and R.E. Rundle, Polynuclear metal carbonyls 1. Structures of $Mn_2(CO)_{10}$ and $Re_2(CO)_{10}$. *J. Chem. Phys.* **1957**, *26*, 1750–1753.

26. R.B. King and M.B. Bisnette, Organometallic chemistry of the transition metals XXI. Some π-pentamethylcyclopentadienyl derivatives of various transition metals. *J. Organomet. Chem.* **1967**, *8*, 287–297.

27. J. Potenza, P. Giordano, D. Mastropaolo, A. Efraty and R.B. King, X-ray crystal structure of dicarbonylpemtamethylcyclopentadienylchromium dimer. *J. Chem. Soc. Chem. Commun.* **1972**, 1333–1334.

28. F.A. Cotton and R.A. Walton, *Multiple Bonds between Metal Atoms*, 2nd ed.; Oxford University Press: Oxford, UK, 1993.

29. A. Nutton and P.M. Maitlis, Pentamethylcyclopentadieny-rhodium and -iridium complexes: XXIII. Di-μ-carbonylbis(pentamethylcyclopentadienylrhodium). *J. Organomet. Chem.* **1979**, *166*, C21–22.

30. S. Takemoto, K. Yoshii, T. Yamano, A. Tsurusaki and H. Matsuzaka, Metal-metal multiple bond formation induced by σ-acceptor Lewis acid ligands. *Chem. Commun.* **2021**, *57*, 923–926.

31. F.A. Cotton, N.F. Curtis, C.B. Harris, B.F.G. Johnson, S.J. Lippard, J.T. Mague, W.R. Robinson and J.S. Wood, Mononuclear and polynuclear chemistry of rhenium(III): its pronounced homophilicity. *Science* **1964**, *145*, 1305–1307.

32. T. Nguyen, A.D. Sutton, M. Brynda, J.C. Fettinger, G.J. Long and P.P. Power, Synthesis of a Stable Compound with Five-Fold Bonding Between Two Chromium (I) Centers. *Science*, **2005**, *310*, 844–847.

33. K.A. Kreisel, G.P.A. Yap, P.O. Dmitrenko, C.R. Landis and K.H. Theopold, The shortest metal-metal bond yet: molecular and electronic structure of a dinuclear chromium diazadiene complex. *J. Am. Chem. Soc.* **2007**, *129*, 14162–14163.

34. A.K. Nair, N.V. Harisomayajula and Y-C. Tsai, The lengths of the metal-to-metal quintuple bonds and reactivity thereof. *Inorg. Chim. Acta* **2015**, *424*, 51–62.

35. N.V. Harisomayajula, A.K. Nair, and Y-C. Tsai, Discovering complexes containing a metal-metal quintuple bond: from theory to practice. *Chem. Commun.* **2014**, *50*, 3391–3411.

36. S. Knecht, H.J.A. Jensen and T. Saue, Relativistic quantum chemical calculations show that the uranium molecule U_2 has a quadruple bond. *Nature Chem.* **2019**, *11*, 40–44.

37. H.K. Sharma, F. Cervantes-Lee, L. Párkány and K.H. Pannell. Catenated Group 14 compounds. Synthesis, structural and spectral characterisation, and chemistry of the chains $Me_3CSiMe_2GeMe_2SnR_3$ (R = Me, Ph). *Organometallics* **1996**, *15*, 429–435.

38. L.W. Reeves, Absolute correlations of nuclear spin-spin coupling constants with atomic number. III. J_{X-H} in isoelectronic series and the lighter elements. *J. Chem. Phys.* **1964**, *40*, 2132–2134.

39. J. Elguero, I. Alkorba and J.E. Del Bene, Calculated coupling constants ^1J(X-Y) and fundamental relationships among reduced coupling constants for molecules H_mX-YH_n, with X,Y = ^{1}H, ^{7}Li, ^{9}Be, ^{11}B, ^{13}C, ^{15}N, ^{17}O, ^{19}F, ^{31}P, ^{33}S and ^{35}Cl. *Magn. Reson. Chem.* **2020**, *55*, 727–732, and references therein.

Chapter 7
Molecules of Very High Symmetry

"Symmetry is what you see at a glance."
— *Blaise Pascal*

7.1. Platonic Polyhedra and the Euler Relationship

The Platonic solids have long fascinated geometers, artists and chemists alike. Only five convex polyhedra, the tetrahedron, cube, octahedron, icosahedron and pentagonal dodecahedron, satisfy the criteria whereby in each case they possess identical regular faces, edges and vertex connectivity. They were discussed in Plato's *Timaeus* (ca. 350 BC) in terms of their supposed association with the classical elements: earth (cube), air (octahedron), fire (tetrahedron) and water (icosahedron), while the fifth element (the dodecahedron) was used to "assign the constellations on the whole heaven". Their mathematical properties were described in the last book of Euclid's *Elements*.

Molecular analogues of the tetrahedron (P_4, B_4Cl_4, $Si_4{}^tBu_4$), octahedron ($[B_6H_6]^{2-}$), cube (C_8H_8), icosahedron ($[B_{12}H_{12}]^{2-}$) and pentagonal dodecahedron ($C_{20}H_{20}$) are now known, and are depicted in Figure 7.1. The tetravalence of carbon makes the C_nH_n molecules viable only for the tetrahedron, cube and dodecahedron, whereas the octahedron and icosahedron are represented by boranes.

It has been recognised for millennia that there is a simple relationship between pairs of Platonic solids. If the centres of adjacent faces of

203

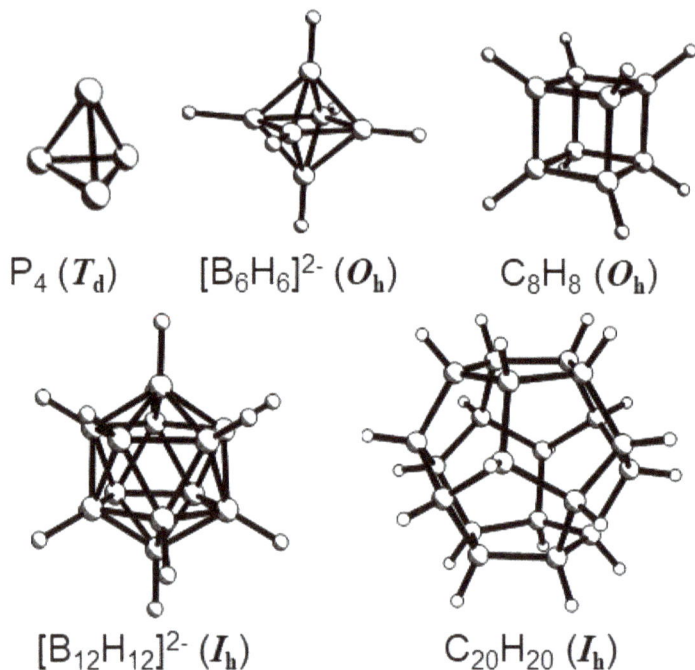

P_4 (T_d)   [B_6H_6]$^{2-}$ (O_h) C_8H_8 (O_h)

[$B_{12}H_{12}$]$^{2-}$ (I_h) $C_{20}H_{20}$ (I_h)

Figure 7.1. Molecular analogues of the Platonic solids.

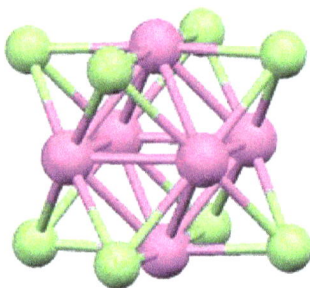

Figure 7.2. The structure of [Mo_6Cl_8]$^{4+}$ demonstrates the reciprocal relationship between the cube and the octahedron.

the octahedron are connected, they yield a cube, and vice versa; this is beautifully illustrated by the X-ray crystal structure of the [Mo_6Cl_8]$^{4+}$ cluster in which an octahedron of molybdenum atoms is encapsulated within a cube of chlorines (Figure 7.2). The icosahedron and pentagonal dodecahedron are similarly related; these pairs of "*recipro-cal*" or "*dual*" polyhedra possess the same point group symmetry.

As noted by René Déscartes around 1620, and stated formally by Leonhard Euler in 1752, for any convex polyhedron there is a simple relationship between the number of vertices (V), faces (F) and edges (E):

$$V + F = E + 2$$

Thus, the cube has 8 vertices, 12 edges and 6 faces; its reciprocal polyhedron — the octahedron — possesses 6 vertices, 12 edges and 8 faces. Likewise, the V, E and F values for the icosahedron (12, 30, 20) and pentagonal dodecahedron (20, 30, 12) are in accord with Euler's equation. Interestingly, the tetrahedron (4 vertices, 6 edges and 4 faces) is its own reciprocal.

7.2. Boranes, Hydrocarbons and Inverse Polyhedra

Closo-borane dianions, and their carborane analogues, adopt polyhedral structures in which each face is triangular; Figure 7.3 shows the *deltahedra* corresponding to the $[B_xH_x]^{2-}$, (x = 5 through 12) or $C_2B_{x-2}H_x$ series, and Table 7.1 lists their point groups and V, E and F values.[1] (One must emphasise that, in these formally electron-deficient systems, the edges do not represent two-electron bonds but merely indicate the structure.) Now, every deltahedron has a reciprocal polyhedron in which each *triangular* face has become a vertex linked to *three* neighbours; this is precisely the criterion to be satisfied for alkanes of the C_nH_n type.

To illustrate this inverse polyhedral relationship between *closo*-boranes, $[B_xH_x]^{2-}$, x = 5–12, and C_nH_n hydrocarbons (n = 6, 8, 10, ..., 20) the point groups and V, E, F values of the complementary C_nH_n molecules are also collected in Table 7.1.

The inverse geometric structures of the *closo*-boranes and their cage hydrocarbon complementary counterparts are rooted in the different electronic configurations of boron and carbon. The C_nH_n systems are assembled from CH units, each of which supplies three atomic orbitals and three electrons to the cage. Each carbon can link to three others via conventional two-electron bonds, thus forming electron-precise molecules. In contrast, BH units also provide three

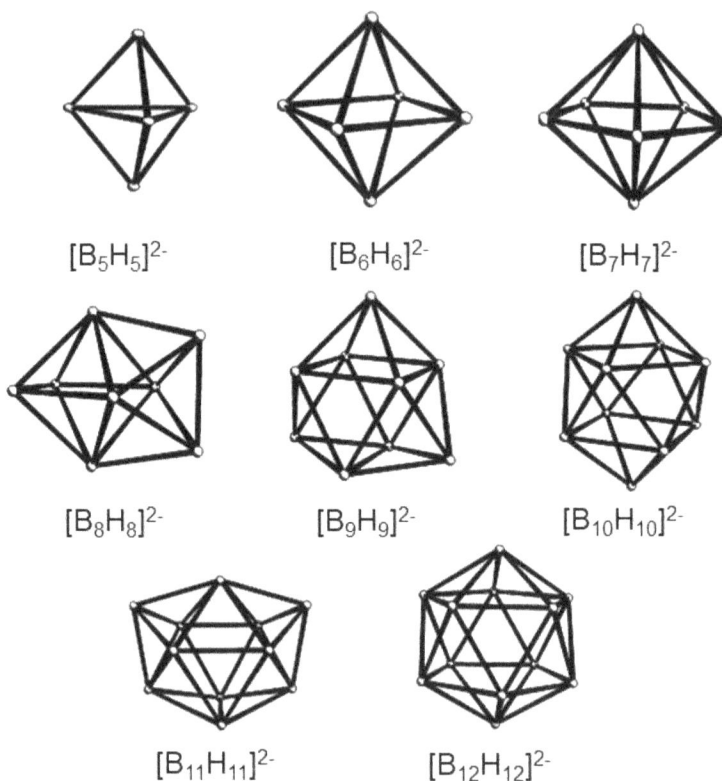

Figure 7.3. The deltahedra corresponding to the structures of the *closo*-boranes $[B_xH_x]^{2-}$.

Table 7.1. Corresponding *closo*-boranes and polycycloalkanes of the same symmetry.

Closo-borane	*V E F*	Point group	*V E F*	Polycycloalkane
$[B_5H_5]^{2-}$	5 9 6	D_{3h}	6 9 5	C_6H_6 [3]prismane
$[B_6H_6]^{2-}$	6 12 8	O_h	8 12 6	C_8H_8 [4]prismane (cubane)
$[B_7H_7]^{2-}$	7 15 10	D_{5h}	10 15 7	$C_{10}H_{10}$ [5]prismane
$[B_8H_8]^{2-}$	8 18 12	D_{2d}	12 18 8	$C_{12}H_{12}$ [$4^4.5^4$]octahedrane
$[B_9H_9]^{2-}$	9 21 14	D_{3h}	14 21 9	$C_{14}H_{14}$ [$4^3.5^6$]nonahedrane
$[B_{10}H_{10}]^{2-}$	10 24 16	D_{4d}	16 24 10	$C_{16}H_{16}$ [$4^2.5^8$]decahedrane
$[B_{11}H_{11}]^{2-}$	11 27 18	C_{2v}	18 27 11	$C_{18}H_{18}$ [$4^2.5^8.6$]undecahedrane
$[B_{12}H_{12}]^{2-}$	12 30 20	I_h	20 30 12	$C_{20}H_{20}$ [5^{12}]dodecahedrane

atomic orbitals but only two electrons for cage bonding; as a result, the *closo*-boranes are electron-deficient molecules with skeletal connectivities greater than three. Their total number of skeletal electron pairs equals the number of vertices plus one; for example, $[B_6H_6]^{2-}$ has 12 electrons in 6 B–H bonds, 7 skeletal electron pairs and is three-dimensionally aromatic. In contrast, in the C_nH_n cages the number of skeletal electron pairs equals the number of edges. In terms of the Euler equation ($V + F = E + 2$), for cage hydrocarbons, $2E = 3V$, as exemplified by cubane, C_8H_8, which has 12 edges and 8 vertices, whereas for *closo*-boranes it is evident that $2E = 3F$, as in $[B_8H_8]^{2-}$, which has 18 edges and 12 faces.

However, one must not assume that bonds are fragile in molecules for which the ratio of valence electrons to interatomic linkages is less than two. For example, the carborane $1,12\text{-}C_2B_{10}H_{12}$ (an icosahedral molecule in which the carbons are maximally separated) only suffers serious decomposition at 630°C, a temperature very much higher than that at which the vast majority of electron-precise organic molecules would survive.

The existence of a complete set of *closo*-boranes, B_xH_{x+2}, or their corresponding anions $[B_xH_x]^{2-}$, where $x = 5$ through 12, suggests that their complementary hydrocarbon cages C_nH_n, where n represents the even numbers 4 through 20, should also all be viable.

7.3. Syntheses of Molecular Platonic Solids and Related Polyhedral Species

7.3.1. *Towards Tetrahedrane*

The search for tetrahedrane has a long history, and the parent molecule still resists isolation. Inevitably, the formation of a molecule containing four mutually connected cyclopropyl moieties clearly introduces considerable additional ring strain. However, Günther Maier and his co-workers in Giessen, Germany were able to prepare the tetra-*tert*-butyl derivative as the first known derivative of this simplest of the Platonic species. As depicted in Scheme 7.1, photolysis of tetra-*tert*-butylcyclopentadienone, **1**, yields initially a "criss-cross" product, **2**, that eventually loses CO to yield tetra-*tert*-butyltetrahedrane, **3**, as

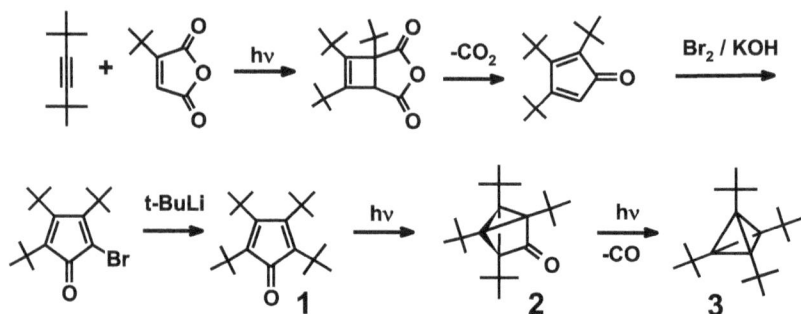

Scheme 7.1.   "Criss-cross" route to tetra-*tert*-butyltetrahedrane.

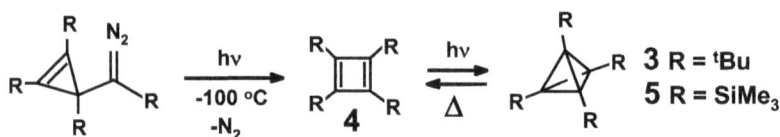

Scheme 7.2.   Cyclopropenyl-diazomethane route to tetra-*tert*-butyltetrahedrane.

a stable crystalline material.[2] However, the difficult multi-step route to tetra-*tert*-butylcyclopentadienone posed major difficulties that were somewhat alleviated by the development of an alternative procedure in which photolysis of the cyclopropenyl-substituted diazomethane furnished tetra(*tert*-butyl)cyclobutadiene, **4** (Scheme 7.2).[3] These isomeric $C_4{}^tBu_4$ species can be interconverted either photochemically (**4→3**) or thermally (**3→4**).[4] This latter approach has also been extended to prepare tetra(trimethylsilyl)tetrahedrane, **5**.

7.3.2. *[3]Prismane, C_6H_6*

The inverse polyhedron to the pentaborane dianion, $[B_5H_5]^{2-}$, is the D_{3h} symmetric cage compound [3]prismane, C_6H_6. Prismane derivatives bearing bulky substituents (e.g., *t*-Bu, CF_3, Ph) have been available for more than three decades via photolysis of sterically encumbered benzenes.[5] The marked deviation from planarity in these systems favours the formation of Dewar benzenes, **6**, which undergo [2+2] cycloadditions to produce the prismane skeleton, **7**, as in Scheme 7.3.

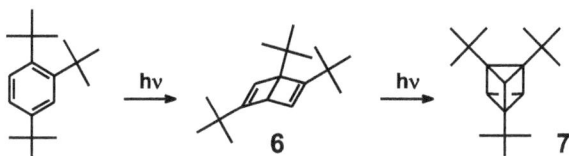

Scheme 7.3. The conversion of Dewar benzenes to [3]prismanes.

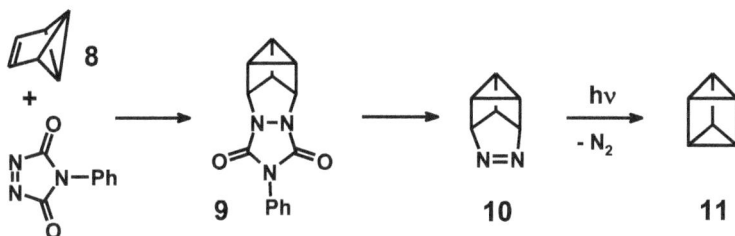

Scheme 7.4. Synthesis of [3]prismane, **11**, by Katz.

However, the parent prismane, **11**, proved much more elusive and was finally obtained in 2% yield by treatment of benzvalene, **8**, with *N*-phenyltriazolinedione to give the cycloadduct **9**; conversion to the azo compound **10** and photolysis to extrude nitrogen finally led to **11** (Scheme 7.4).[6] Although the yield of the final photolysis step has now been improved somewhat to 15%,[7] [3]prismane is still not a conveniently obtainable molecule.

7.3.3. *[4]Prismane, C_8H_8 (Cubane)*

The key step of Philip Eaton's beautiful synthesis of [4]prismane (cubane, **12**),[8] shown in Scheme 7.5, is the double Favorskii ring contraction previously discussed in Chapter 2.

An elegant modification of this procedure has been reported by Pettit who used (cyclobutadiene)Fe(CO)$_3$, **13**, as the source of one of the square faces (Scheme 7.6).[9] Oxidation of the organometallic precursor by cerium(IV) nitrate liberated cyclobutadiene that underwent a [4+2] cycloaddition with 2,5-dibromo-*p*-quinone to form **14** after photolytic ring closure. Once again, as in Eaton's procedure, two

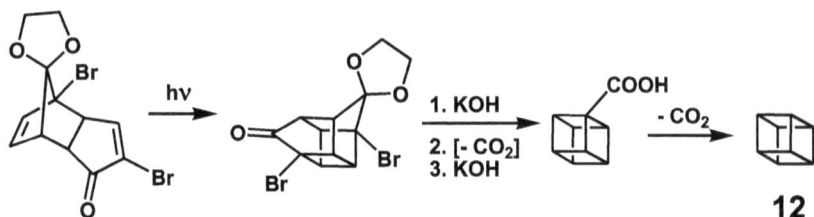

Scheme 7.5.   First synthesis of cubane, **12**, by Eaton.

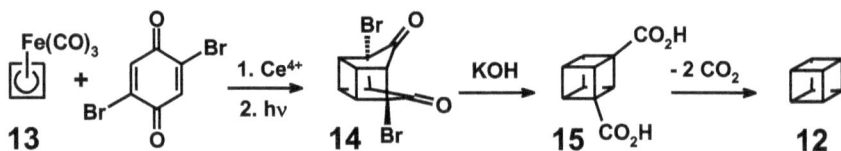

Scheme 7.6.   Synthesis of cubane by Pettit.

Favorskii ring contractions were used to obtain the cubane skeleton, first as the dicarboxylic acid **15**, which is subsequently decarboxylated to yield **12**.

The original approach has since been considerably improved and modified so that functionalised cubanes can now be obtained in kilogram quantities, and their chemistry has been extensively studied.[10] For example, polynitrocubanes have been investigated as high-energy-density materials,[11–13] and the cardiopharmacological activity of cubane dicarboxylic acid and its amide has been reported.[14]

7.3.4. *[5]Prismane, $C_{10}H_{10}$ (Pentaprismane)*

Since pentaprismane, **16**, is the least strained of the prismanes, one might have expected it to be readily available by photolytic [2+2] cycloaddition of hypostrophene, **17**, or by extrusion of nitrogen from either **18** or **19**, as in Scheme 7.7;[15–18] surprisingly, all these routes were ineffective!

Success was finally achieved via ring contraction of a homopentaprismane, somewhat analogous to the original cubane synthesis. As shown in Scheme 7.8, Diels-Alder reaction of 1,2,3,4-tetrachloro-5,5-dimethoxycyclopentadiene with *p*-benzoquinone gave **20**, which

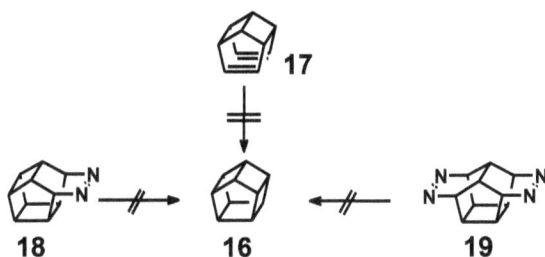

Scheme 7.7. Unsuccessful routes to [5]prismane.

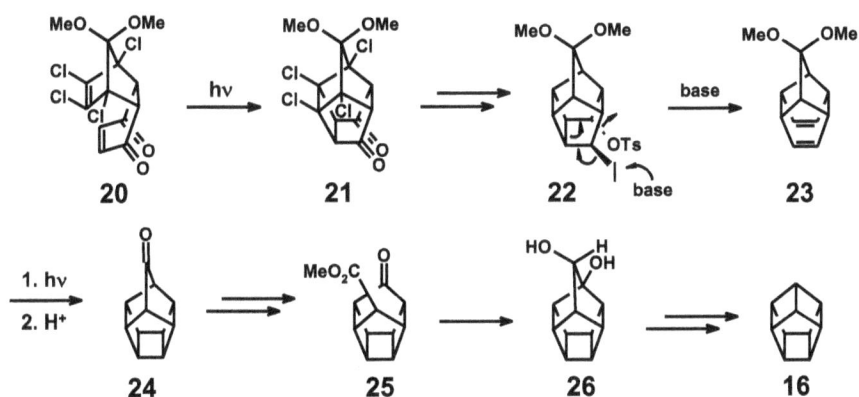

Scheme 7.8. Synthesis of [5]prismane by Eaton.

underwent photolytic [2+2] closure to the pentacyclic dione, **21**. Subsequent dechlorination and functional group manipulation led to the iodo-tosylate, **22**, which, in the presence of base, generated the homo-hypostrophene, **23**; [2+2] cycloaddition then furnished the homopenta-prismanone, **24**. Introduction of a bridgehead bromine (with the intent of carrying out a Favorskii ring contraction) proved impossible. Instead, it was necessary to proceed via the keto-ester **25** and the dihydroxy-homopentaprismane **26** which, after ring contraction and decarboxyla-tion, yielded **16**.[19,20]

7.3.5. *Pentagonal Dodecahedrane, $C_{20}H_{20}$*

The preparation of dodecahedrane, **27**, is undoubtedly one of the great synthetic achievements of recent times, and will remain at the

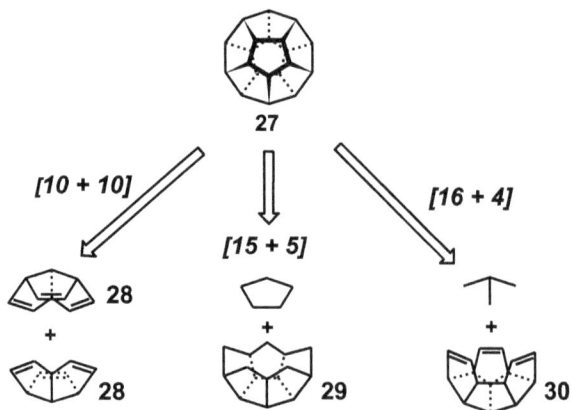

Scheme 7.9.   Retrosynthetic analysis for potential approaches to dodecahedrane.

pinnacle of alicyclic chemistry until a stepwise synthesis of C_{60} is achieved. However, we note *en passant* that Scott, de Meijere and their colleagues have devised a rational route to C_{60} from a chlorinated precursor, $C_{60}H_{27}Cl_3$, in which only the final ring closures proceed via preparative-scale flash vacuum pyrolysis.[21]

The icosahedral point group possesses ten C_3 axes and six C_5 axes, and synthetic proposals (Scheme 7.9) have taken advantage of both types of symmetry element. The former prompted Woodward[22] and Jacobson[23] independently to suggest that two triquinacene units, **28**, could be coupled. A second approach is based upon the five-fold symmetry of [5]peristylane, **29**; this system has been accessed by Eaton but attempts to add the 5-carbon roof have not yet succeeded.[24] Another "3-fold" approach relies on the addition of a trimethylenemethane-like C_4 fragment to C_{16}-hexaquinacene, **30**; again complications arose during attempts to convert this molecule to **27**.[25]

The requisite $C_{10}H_{10}$ moiety triquinacene, **28**, was first prepared by Woodward in a rather lengthy multi-step sequence (Scheme 7.10) that began by conversion of the tetracyclic alcohol, **31**, into the epoxy ketone, **32**. Base-promoted formation of the additionally required third five-membered ring, together with ring opening of the epoxide, gave the keto alcohol, **33**. The process continued via the dihydroxy ether, **34**, and the anhydride, **35**, leading eventually to the *endo,endo*

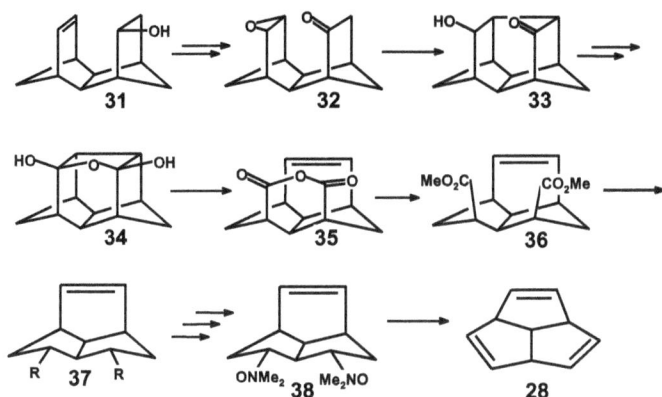

Scheme 7.10. Key steps in R.B. Woodward's route to triquinacene, **28**.

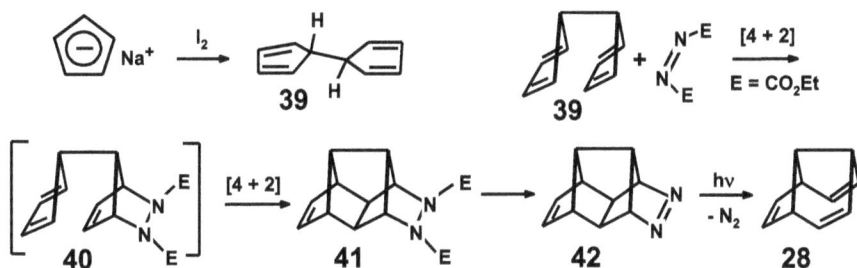

Scheme 7.11. Leo Paquette's route to triquinacene, **28**.

diester, **36**, that was subsequently isomerised into the *exo,exo* diester, **37** R = CO$_2$Me. There then followed a rather long sequence via the corresponding acyl chlorides, acyl azides, isocyanates (Curtius rearrangement), and urethanes, **37** R = NHCO$_2$Me, and eventually the bis amine oxide, **38**. Finally, a double Cope elimination of Me$_2$NOH gave **28**.[22]

This route has since been superseded by a very short synthesis reported by Leo Paquette from Ohio State University in Columbus (Scheme 7.11).[26] By analogy to a magnificently inventive chess game, a grandmaster would describe it as a "brilliancy". The iodine-promoted dimerisation of C$_5$H$_5$Na to form the C$_{10}$H$_{10}$ isomer, **39**, was followed by reaction with diethyl azodicarboxylate initiating a spectacular *domino Diels-Alder* sequence whereby the initially formed [4+2] adduct, **40**, was followed by a second such cycloaddition to generate **41**. Hydrolysis and oxidation with MnO_2 to give the azo compound, **42**, and photolysis to eliminate N_2 delivered triquinacene in good yield. However, all attempts at controlled dimerisation of **28** to form dodecahedrane, even on a transition metal template,[27] have so far proven fruitless.

As outlined in Paquette's eloquent overview of the history of the dodecahedrane project,[28] success was finally achieved via the C_2 route summarised in Scheme 7.12. The crucial intermediate, **43**, was hydrogenated to **44**, which by several cyclisation steps was converted to the diol, **45**. Its oxidation and condensation of the resulting ketoaldehyde then provided the mono ketone, **46**, which was photo-chemically ring-closed to the secododecahedrene, **47**. After its diimine hydrogenation to **48**, only one C–C bond in the nearly completed sphere was lacking. This final cyclodehydrogenation to **27** was accomplished by treatment of **48** with Pd/C at 250°C.[29]

Subsequently, Prinzbach established an entirely different route to dodecahedrane (Scheme 7.13) that proceeds by catalytic isomerisation of pagodane, a molecule of D_{2h} symmetry. Photolysis of **49** yields the bis-(1,3-cyclohexadiene), **50**, and sets up the system for domino Diels-Alder cycloadditions initiated by maleic anhydride to generate **51**. The reaction sequence continues via diene **52**, diol **53**, diketone

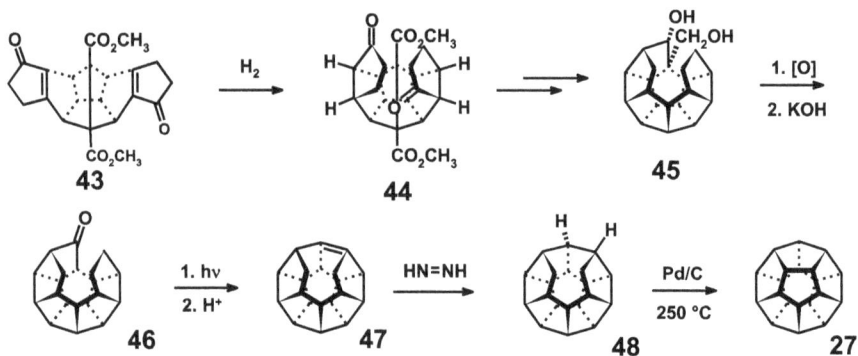

Scheme 7.12. Paquette's synthesis of dodecahedrane, **27**.

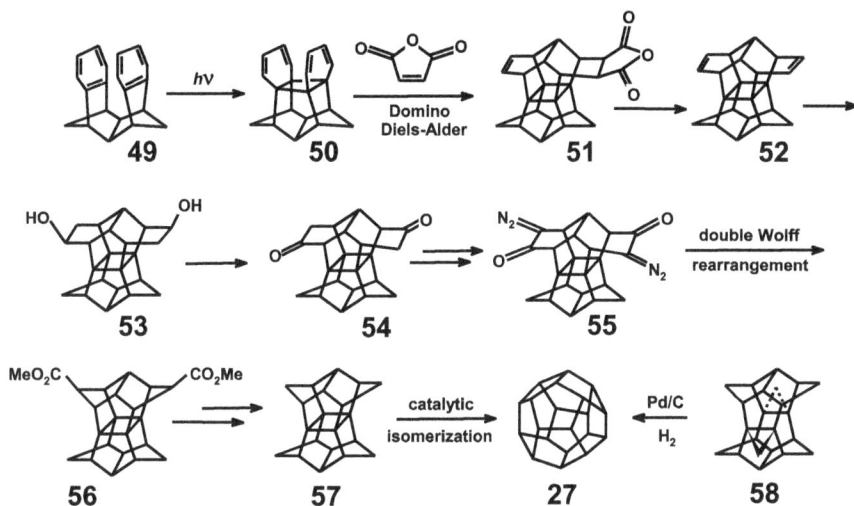

Scheme 7.13.   Selected key steps in Prinzbach's synthesis of dodecahedrane, **27**.

54, and the bis-diazoketone **55**. A double Wolff rearrangement to form diester **56** leads eventually to pagodane, **57**, and finally to dodecahedrane, **27**. Subsequently, the yield of the final ring-closing process was improved whereby the bis-cyclopropanated hydrocarbon **58** yielded dodecahedrane on treatment with Pd/C in a hydrogen atmosphere. Prinzbach's eloquent and comprehensive review of his group's major contributions to this area makes for valuable reading.[30]

7.3.6. *Octahedrane, $C_{12}H_{12}$*

Having very briefly outlined the successful syntheses of several poly-cyclic alkanes, C_nH_n, where n = 4, 6, 8, 10 and 20, we turn now to the remaining members for which n = 12, octahedrane, n = 14, non-ahedrane, n = 16, decahedrane, and n = 18, undecahedrane. Although none of these have yet been reported, there are a number of "near misses", and there is much beautiful chemistry associated with their potential precursors; one can only admire the determination and ingenuity of the investigators in this area.

The octaborane dianion, $[C_8H_8]^{2-}$, exhibits D_{2d} symmetry and one could envisage the preparation of its $C_{12}H_{12}$ counterpart, **59**, by coupling of two Dewar benzenes, as in **60**, but we are unaware of any such reports.

60

However, metal carbonyl-promoted dimerisation of norbornadienes, as in the formation of **61**, is a well-established protocol, and Marchand has exploited this reaction to prepare the C_{14} diketone, **62** (Scheme 7.14).[31] In principle, one could incorporate bridgehead halogens, as in **63**, with the aim of carrying out two Favorskii ring contractions to generate the octahedrane skeleton as the dicarboxylic acid, **64**, which ultimately would have to be decarboxylated to the parent system, **59**. More realistically, one can see the obvious similarity to the conversion of homopenta-prismanone, **24**, to penta-prismane, **16**, which was successfully accomplished via Bayer-Villiger oxidation, acyloin coupling and decarboxylation. Thus, one might anticipate that such an approach might provide access to D_{2d}-$[4^4.5^4]$ octahedrane, **59**. (We note that a simplified nomenclature has been

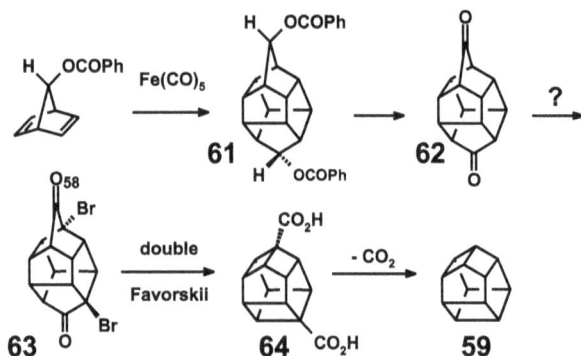

Scheme 7.14.   A possible route to D_{2d} octahedrane, **59**.

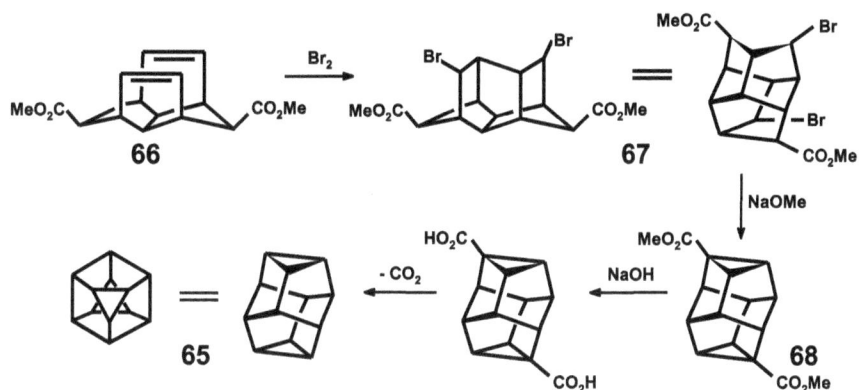

Scheme 7.15. De Meijere's route to D_{3d} octahedrane, **65**.

proposed in which the number of 3, 4 or 5-membered rings is indicated by a superscript; thus cubane is $[4^6]$hexahedrane and pentaprismane is $[4^5.5^2]$heptahedrane.[32])

However, de Meijere has reported that the energetically most favoured $(CH)_{12}$ structure is the D_{3d}-symmetric isomer p-$[3^2.5^6]$octahedrane, **65**. As depicted in Scheme 7.15, bromination of the diester of the tetracyclic diene, **66**, yielded the dibromo compound **67** which, when treated with NaOMe, eliminates HBr twice and yields the desired molecular skeleton possessing two new cyclopropyl rings, as in **68**. Hydrolysis and decarboxylation delivered **65** that was fully characterised spectroscopically and by X-ray crystallography.[33]

An entirely different route to the D_{3d} octahedrane framework arose from a study of the photolytic behaviour of a series of 2,11-diaza[3.3]paracyclophanes. After irradiation of cyclophane, **69**, at 300 nm for 36 h, repeated chromatographic separation gave a 33% yield of the octahedrane, **70**, bearing two three-atom bridges, whose structure was confirmed by X-ray crystallography.[34] As indicated in Scheme 7.16, coupling between the benzene rings leads to the formation of two cyclohexane-like chair fragments that also contain cyclopropyl rings.

Yet another formal dimer of benzene has been prepared by the reaction of the bishomocubanone, **71**, with diazomethane. The resulting ketone was treated with tosylhydrazine, and the product

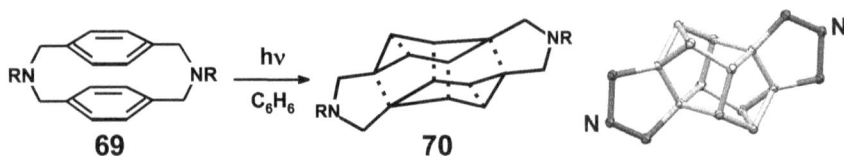

Scheme 7.16. Photochemical rearrangement of a 2,11-diaza[3.3]paracyclophane, **69**, into a double-bridged octahedrane, **70**, and the X-ray structure showing its molecular framework.

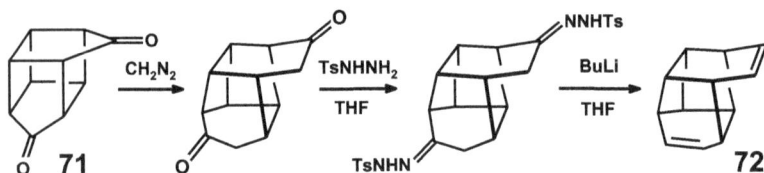

Scheme 7.17. Formation of an *o,o':o,p'* benzene dimer, $C_{12}H_{12}$.

underwent base-promoted Shapiro reactions to yield C_2-symmetric **72**, in which the benzene rings are now cross-linked in an *o,o':o,p'* fashion (Scheme 7.17).[35]

7.3.7. *Nonahedrane*, $C_{14}H_{14}$

It has been proposed that an intermediate in the synthesis of [5]peristylane may be a viable precursor to the still-unknown D_{3h}-[$4^3.5^6$]-nonahedrane, **73**. As shown in Scheme 7.18, the diene-dione, **74**, can be readily converted to the double enone, **75**, which undergoes [2+2]photocyclisation to the pentacyclic diketone, **76**.[36] We await further elaboration of this fascinating system.

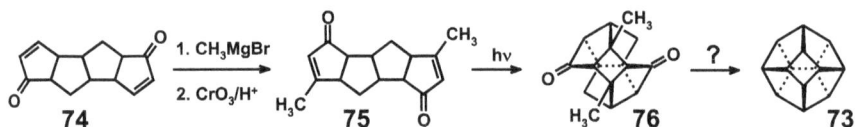

Scheme 7.18. A possible route to D_{3h} nonahedrane, **73**.

7.3.8. *Decahedrane, $C_{16}H_{16}$*

Formally, the D_{4d} decahedrane, **77**, should be available by capping [4] peristylane with a four-membered ring system, as in **78**. However, to translate this concept into a preparatively realistic protocol is a different matter and, although there is no report of a completed synthesis of **77**, considerable progress has been made.[37]

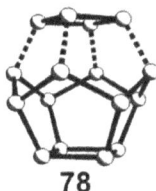

78

The route taken by Paquette and co-workers (Scheme 7.19) required the initial generation of the fulvene **79**, whose four-membered ring should eventually serve as the "roof" of a [4]-peristylane "building". Towards this goal, **79** was first converted into the cyclopentadiene derivative **80**, whose carbon skeleton was subsequently extended and then bent into a convex shape by an epoxidation reaction, thus forming **81**. After the still saturated C_2-bridge had been converted to an etheno bridge, the prerequisite for an intramolecular [2+2] photoaddition had been created. Indeed, photochemical ring closure and various oxidation steps next led to the "open" triketone **82** that in principle should be convertible by two aldol condensations to a seco-decahedrane skeleton; this latter system might then be persuaded to close to **77** by taking advantage of methodology established during the dodecahedrane project.

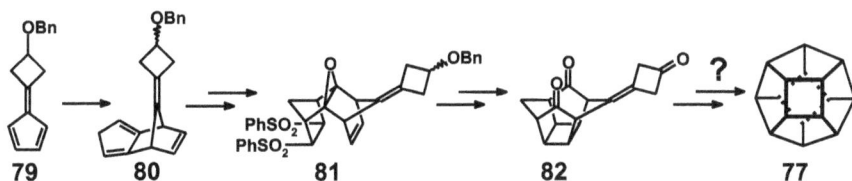

Scheme 7.19. A possible route to D_{4d} decahedrane.

Finally, we note that the reciprocal polyhedron to the octadecahedron $[B_{11}H_{11}]^{2-}$ is C_{2v}-symmetric $[4^2.5^8.6]$-undecahedrane, **83**. Note that this $C_{18}H_{18}$ system would contain a six-membered ring paralleling the C_{2v} symmetry of the borane that has a capping boron linked to six others. However, we are unaware of any attempted syntheses of this molecule.

7.4. Buckminsterfullerene, C_{60}

The initial report[38] of the detection of C_{60} upon laser irradiation of carbon led to an avalanche of publications, and Nobel Prizes for Kroto, Smalley and Curl. It was verified by X-ray crystallography that this molecule is an icosahedral polyhedron with 32 faces, 12 of them pentagonal and the other 20 hexagonal.[39] The trivial name is derived from that of the architect Buckminster Fuller who famously designed geodesic dome structures, but is now perhaps more commonly associated with the football motif (buckyballs). In terms of its Euler formulation, the *V, E, F* values are 60, 90, 32; the system possesses 240 valence electrons of which 180 are required for the 90 edge linkages leaving 60 for multiple bonding, whereby each carbon is involved in one double and two single bonds, although very many canonical structures can be drawn. Despite the development of the now widely used route to C_{60} by resistive heating of graphite,[40] that has made it currently available commercially in 99.99% purity at ~\$330/g, efforts to establish a more logical stepwise procedure are continuing. The original authors were probably unaware at the time of the long-running project in Orville Chapman's group at UCLA towards a designed synthetic route to C_{60}. Current major focus is on the preparation and chemistry of important subunits of C_{60}, such as *semi-buckminster-fullerene*, $C_{30}H_{12}$, *corannulene*, $C_{20}H_{10}$, and *sumanene* $C_{21}H_{12}$.

The three-fold symmetry of C_{60} has prompted numerous studies on the trimerisation of potential subunits. Typically, acid-catalysed cyclisation of cyclopentanone or indanone yields trindane, **84**, or truxene, **85**, both of whose chemistry have been exploited.[41–43] The climax of such an approach has seen a designed stepwise synthesis of C_{60}, albeit in very low yield. Cyclisation of the ketone **86** led to **87** that was selected with the goal of attempting multiple dehydrohalogenations and dehydrogenations so as to bring about a "stitching together" of the three arms of the molecule via flash vacuum pyrolysis (FVP), thus forming C_{60}. Gratifyingly, this strategy (Scheme 7.20) was successful, and the only fullerene produced was C_{60} that was unambiguously characterised by mass spectrometry.[21]

A number of syntheses of molecules representing sizeable portions of the C_{60} skeleton have been prepared, notably by Peter Rabideau at Louisiana State University, who reported the syntheses of the semibuckminsterfullerenes $C_{30}H_{12}$, **88** and **89**.[44] However, since the final step involved FVP, the available quantities did not permit extensive investigation of their chemistry. Subsequently, however, it was found that the FVP ring-closing step could be avoided by using low-valent vanadium or titanium reagents to bring about coupling via multiple debrominations.[45]

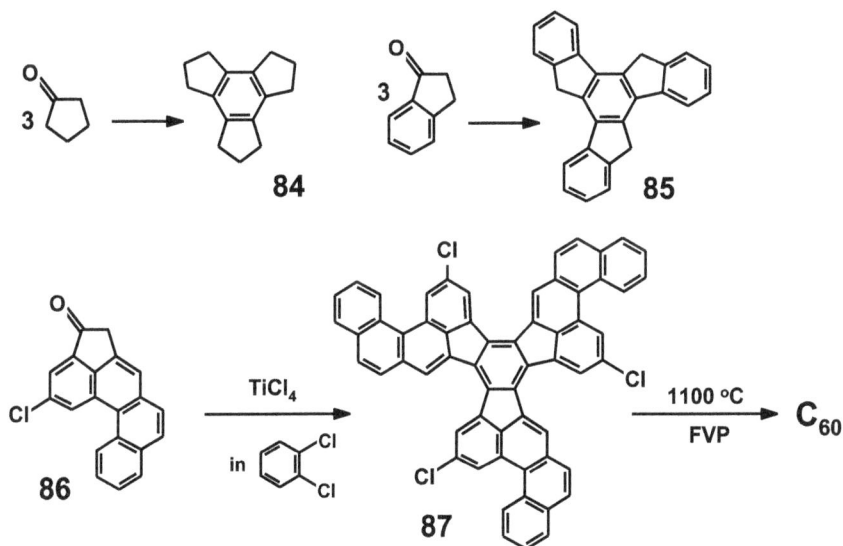

Scheme 7.20. A rational synthesis of C_{60}.

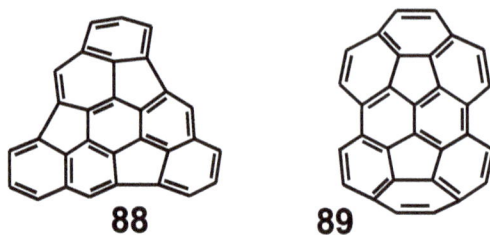

88   **89**

In other work, the Schmittel cyclisation of an *in situ* generated benzo-enyne-allene, **90**, provided a route to the nonacyclic 34-carbon system, **91**, that represents a major portion of the C_{60} skeletal framework (Scheme 7.21).[46]

Much recent attention has been focussed on the availability of corannulene, **92**, and sumanene, **93**, and their current availability in bulk quantities has allowed very extensive development of their reactivity.

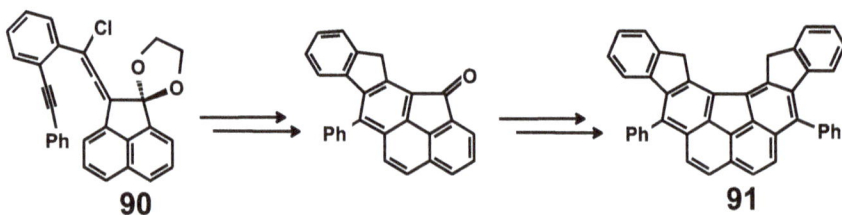

90   **91**

Scheme 7.21.   Schmittel's route to a nonacyclic fragment of C_{60}.

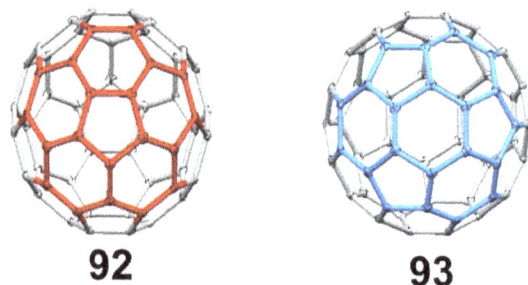

92   **93**

Figure 7.4.   Corannulene, **92**, and sumanene, **93**, mapped in colour onto the C_{60} framework.

7.4.1. *Corannulene, $C_{20}H_{10}$*

Corannulene has a central five-membered ring surrounded by six-membered rings. The first synthesis was reported by Lawton and Barth in 1966;[47] it was a difficult 17-step procedure with a 1% overall yield, and has since been very greatly improved, mostly by the efforts of Larry Scott at the University of Nevada, and by Jay Siegel in San Diego and later in Zürich.

Scott's original route involved a Knoevenagel condensation to generate the cyclopentadienone, **94**, that underwent a Diels-Alder cycloaddition with norbornadiene to form **95**; this is exquisitely poised to suffer a retro-Diels-Alder elimination of cyclopentadiene and cheletropic loss of CO to deliver the tetracycle, **96**, that can possess a range of different substituents. In fact, these can be telescoped into a single step when carried out at 120°C using norbornadiene as the solvent (Scheme 7.22). The overall effect of the two Diels-Alder steps is to deliver an acetylene unit, thereby forming the third aromatic ring. When the new substituent is an ethynyl group, as in **97**, under pyrolytic conditions (FVP 1000°C) it rearranges, presumably stepwise, to form a vinylidene, **98**, and this transient carbene can insert into aromatic C–H bonds to furnish corannulene in ~10% yield (Scheme 7.22).[48]

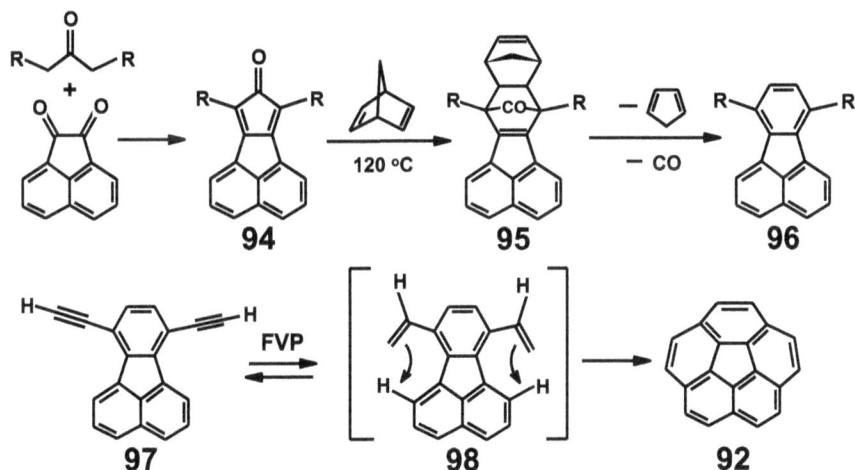

Scheme 7.22. Scott's original route to corannulene.

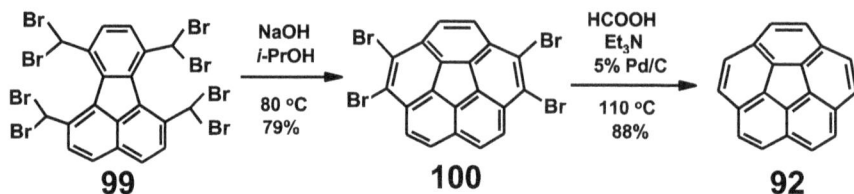

Scheme 7.23. Key steps in Siegel's kilogram-scale synthesis of corannulene.

The history of the gradual improvements in the process has been chronicled by Siegel,[49,50] who has now taken it to the kilogram level — an enormously praiseworthy achievement! In this large-scale preparation, the final steps involve formation of the tetrakis(dibromomethyl) derivative, **99**, ring closure to the tetrabromide **100**, under conventional laboratory conditions (80°C in isopropanol), and finally debromination to form corannulene in excellent yield (Scheme 7.23). At this point, corannulene has become a readily available material and its chemistry has since been extensively investigated.[51]

7.4.2. Sumanene, $C_{21}H_{12}$

Sumanene has a central six-membered ring surrounded by alternating five- and six-membered rings. The name is derived from the Sanskrit word *suman* which translates as "flower", and the ring edges are considered to resemble petals. An early attempt at its synthesis by Mehta involved FVP of 1,5,9-trimethyltriphenylene, **101**, or tris(bromomethyl)-triphenylene, **102**, but instead yielded only the singly or doubly bridged products, **103** and **104**, respectively (Scheme 7.24).[52]

Success was eventually achieved by Sakurai, Daiko and Hirao who prepared a bromo-lithio derivative of norbornadiene, **105**, that was converted into the corresponding tin compound, **106** (Scheme 7.25). Copper-mediated trimerisation of **106** gave a 3:1 mixture of *anti* and *syn* benzotris(norbornadiene), **107**, the latter of which underwent metathesis with the Grubbs catalyst to furnish hexahydro-sumanene, **108**. The process was completed upon oxidation with DDQ to deliver sumanene, **109**, with no need for an FVP step.[53]

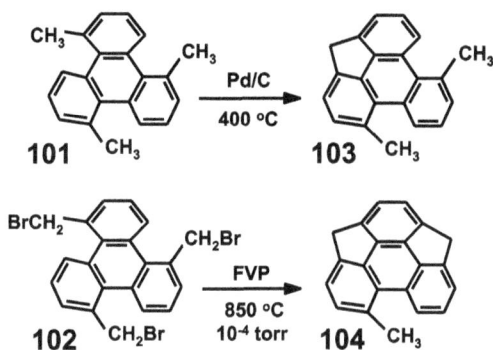

Scheme 7.24. Mehta's pioneering work on a potential route to sumanene.

Since that time, the chemistry of sumanene has expanded almost exponentially to include heterosumanenes, typified by **110**, as well as numerous polysumanenes and multiply substituted sumanenes. This work has been comprehensively reviewed recently.[54,55]

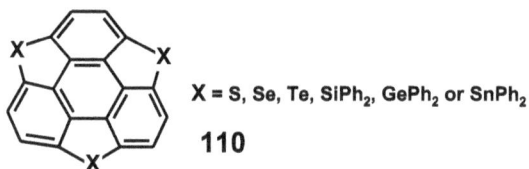

X = S, Se, Te, SiPh$_2$, GePh$_2$ or SnPh$_2$

110

7.5. Highly Symmetric Inorganic Polyhedranes

The major synthetic challenges that needed to be overcome to synthesise the polyhedranes [3]prismane, **11**, cubane, **12**, and [5]prismane, **16**, stand in stark contrast to the ready availability of their heavier congeners. Thus, trigonal prismatic hexasilanes and hexagermanes, **111**, are preparable in single-step processes by sodium- or magnesium-mediated dehalogenation of the appropriate REX$_3$ precursor, where R is a very bulky alkyl or aryl group, and X is chlorine or bromine (Scheme 7.26).[56,57] The hexatellurium cation, $[Te_6]^{4+}$, is also trigonal prismatic.[58,59]

Similarly, the inorganic cubane analogues R$_8$E$_8$, where E = Si or Ge, **112**, and R is again a bulky group, are also well known (Scheme 7.27);

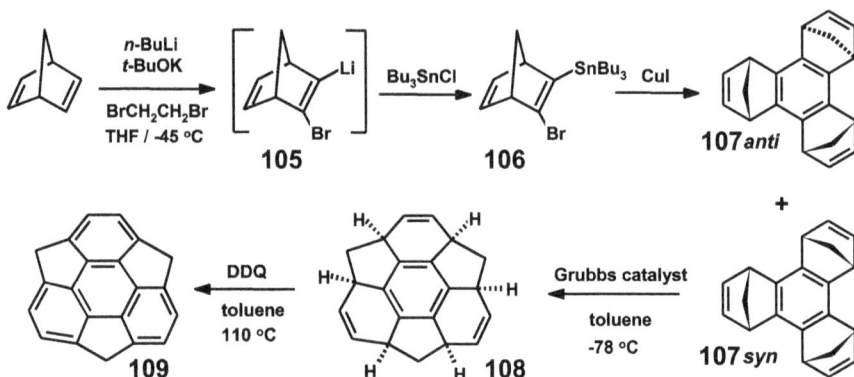

Scheme 7.25. The Sakurai-Daiko-Hirao route to sumanene, **109**.

Scheme 7.26. Synthetic routes to trigonal prismatic hexasilanes and hexagermanes.

Scheme 7.27. Synthetic routes to octasila- and octagerma-cubanes.

indeed, octakis(*t*-butyldimethylsilyl)octasilane is preparable in 72% yield by treatment of the corresponding trichlorosilane precursor with sodium in toluene at 90°C. These systems have been thoroughly investigated structurally and spectroscopically, and their reactivity has

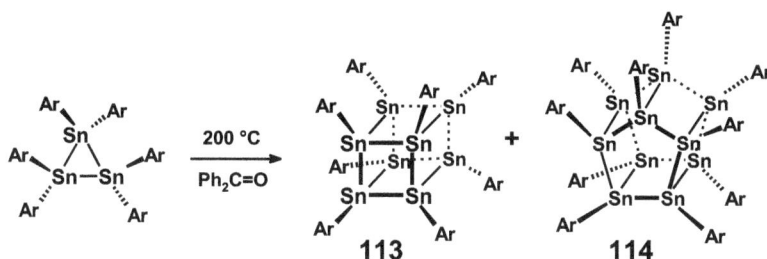

Scheme 7.28. Synthesis of an octastannacubane and a decastannapentaprismane.

also been extensively investigated.[60–62] Furthermore, as shown in Scheme 7.28, the octastannacubane, **113**, and the per-arylated decastannane, **114**, a tin analogue of pentaprismane, **11**, have been prepared by thermolysis of hexakis(2,6-diethylphenyl)cyclotristannne and fully characterised spectroscopically and by X-ray crystallography.[63–65]

The major mitigating factor here is that such elements commonly form structures in which 90° angles are the norm,[66] and so ring strain is no longer such a major impediment to bond formation. While the intrinsic yields of these products can range from very good to rather poor, this is compensated by the fact that they involve short syntheses from relatively inexpensive starting materials.

We should also mention the structures of a series of main group cluster ions, originally recognised by Eduard Zintl (1898–1941) at the University of Freiburg in Germany, and now named in his honour as Zintl ions. Many of these are best prepared by reaction of elements such as germanium, lead or antimony in liquid ammonia solution. Because of their sensitivity, they are generally isolated by using a cation-sequestering agent, as in $[K(2,2,2\text{-crypt})]_3[Sb_7]$, thereby allowing them to be characterised by X-ray crystallography. In many cases, their structures correspond to those seen in the borane anions, $[B_xH_x]^{2-}$, depicted in Figure 7.3. Thus, $[Sn_5]^{2-}$, $[Pb_9]^{2-}$, $[Ge_{10}]^{2-}$, and $[Sn_{12}]^{2-}$ are trigonal bipyramidal, D_{3h}, tricapped trigonal prismatic, D_{3h}, bicapped square antiprismatic, D_{4d}, and icosahedral, I_h, respectively. This area has been comprehensively reviewed recently.[67] Their bonding follows the pattern established in the Williams-Wade-Mingos-Rudolph Polyhedral Skeletal Electron Pair Theory (PSEPT) whereby each element in the cluster bears an external lone pair and the total number of electron pairs involved in skeletal bonding is $N + 1$ for an N-vertex *closo* polyhedron.[68] For example, the $[Pb_9]^{2-}$ cluster is a 38-electron system; 18 electrons are required for the nine external lone pairs, and the remaining 20 electrons (10 pairs) correspond to the 9-vertex tricapped trigonal prismatic structure.

In addition, a large number of transition metal clusters exhibiting very high symmetry are known. Typically, the $Ni_6(C_5H_5)_6$ is octahedral, and in $[Au_{13}Cl_2(PR_3)_{10}]^{3+}$ twelve of the gold atoms form an icosahedron with the 13th gold sited interstitially within the cage.[69] We note also that King *et al.* have pointed out the reciprocal polyhedral relationship between gold clusters and fullerenes.[70]

7.6. Closing Remarks

It is evident that there exists a complementary relationship between *closo*-boranes $[B_xH_x]^{2-}$, where $x = 5$ through 12, and polycycloalkanes C_nH_n, where n represents the even numbers from 6 through 20. Several of these hydrocarbons are known while others remain elusive. Interestingly, one can invert the original concept and propose that other highly symmetrical cage hydrocarbons of the C_nH_n type might have *closo*-borane counterparts. In particular, it has been proposed that C_{60} (V, E, F = 60, 90, 32) could have a corresponding *closo*-borane $[B_{32}H_{32}]^{2-}$ (V, E, F = 32, 90, 60) of icosahedral symmetry.[71] Moreover, Lipscomb and Massa discussed the structures of borane analogues of fullerenes,[72,73] and even of nanotubes;[74] indeed, boron nitride nanotubes are now commercially available.

Molecular analogues of several members of both sets of the complementary polyhedra exhibited by the *closo*-boranes and by the polycycloalkanes have been constructed from other elemental species: clusters containing lithium, transition metals, silicon, phosphorus, arsenic, bismuth, lead, etc., have been characterised, and their architectures continue to delight us. The existence of molecules of such exquisite symmetry would surely have fascinated Plato.

References

1. M.J. McGlinchey and H. Hopf, Reciprocal polyhedra and the Euler relationship: cage hydrocarbons, C_nH_n and *closo* boranes $[B_xH_x]^{2-}$. *Beilstein J. Org. Chem.* **2011**, *7*, 222–233.
2. G. Maier, S. Pfriem, U. Schäfer and R. Matusch, Tetra-*tert*-butyltetrahedrane. *Angew. Chem. Int. Ed. Engl.* **1978**, *17*, 520–521.
3. G. Maier and F. Fleischer, An alternative route to tetra-*tert*-butyltetrahedrane. *Tetrahedron Lett.* **1991**, *32*, 57–70.
4. G. Maier, J. Neudert, O. Wolf, D. Pappusch, A. Sekiguchi, M. Tanaka and T. Matsuo, Tetrakis(trimethylsilyl)tetrahedrane. *J. Am. Chem. Soc.* **2002**, *124*, 13819–13826.
5. G. Mehta and S. Padma, Syntheses of Prismanes. In *Carbocyclic Cage Compounds*, Wiley-VCH: Weinheim, Germany, 1992; pp. 189–191.
6. T.J. Katz and N. Acton, Synthesis of prismane. *J. Am. Chem. Soc.* **1973**, *95*, 2738–2739.
7. N. Turro, V. Ramamurthy and T.J. Katz, Energy-storage and release — direct and sensitized photoreactions of Dewar benzene and prismane. *Nouv. J. Chim.* **1977**, *1*, 363–365.
8. P.E. Eaton and T.W. Cole Jr., Cubane. *J. Am. Chem. Soc.* **1964**, *86*, 3157–3158.
9. J.C. Barborak, L. Watts and R. Pettit, A convenient synthesis of the cubane system. *J. Am. Chem. Soc.* **1966**, *86*, 1328–1329.
10. G.W. Griffin and A.P Marchand, Synthesis and chemistry of cubanes. *Chem. Rev.* **1989**, *89*, 997–1010.
11. P.E. Eaton, B.K. Ravi Shankar, G.D. Price, J.J. Pluth, E.E. Gilbert, J. Alster and O. Sandus, Synthesis of 1,4-dinitrocubane. *J. Org. Chem.* **1984**, *49*, 185–186.
12. A. Bashir-Hashemi, S. Iyer, J. Alster and N. Slagg, Cubanes and cage related molecules. *Chem. Ind.* **1955**, 551–555.
13. M.-X. Zhang, P.E. Eaton and R. Gilardi, Hepta- and octanitrocubanes. *Angew. Chem. Int. Ed.* **2000**, *39*, 401–404.
14. L.T. Eremenko, L.B. Romanova, M.E. Ivanova, D.A. Nesterenko, V. Malygina, A.B. Eremeev, G.V. Logodzinskaya and V.P. Lodygina, Cubane derivatives 3. Synthesis and antiischemic activity of some nitroxyalkyl derivatives of 1,4-cubanedicarboxylic acid and its diamide. *Russ. Chem. Bull.* **1998**, *47*, 1137–1140.
15. K.-W. Shen, Synthesis and reactions of 2,3,7,8-tetraazahexacyclo[7.4.1^{04},12^{05},-14^{06},11.0^{10},13-tetradeca-2,7-diene. *J. Am. Chem. Soc.* **1971**, *93*, 3064–3066.
16. E.L. Allred and B.R. Beck, 10,11-Diazahexacyclo[6.4.0.0^{2,7}.O^{3,6}.O^{4,12}.O^{5,9}] dodec-10-ene. A new precursor for $(CH)_{10}$ compounds. *Tetrahedron Lett.* **1974**, *15*, 437–440.
17. R. Pettit, J.S. McKennis, L. Brener and J.S. Ward, Degenerate Cope rearrangements in hypostrophene, a novel $C_{10}H_{10}$ hydrocarbon. *J. Am. Chem. Soc.* **1971**, *93*, 4957–4958.

18. L.A. Paquette, R.F. Davis and D.R. James, A simple route from cyclopentadiene to hypostrophene. *Tetrahedron Lett.* **1974**, *15*, 1615–1618.

19. P.E. Eaton, Y.S. Or and S.J. Branka, Pentaprismane. *J. Am. Chem. Soc.* **1981**, *103*, 2134–2136.

20. P.E. Eaton, Y.S. Or, S.J. Branka and B.K. Ravi Shankar, The synthesis of pentaprismane. *Tetrahedron* **1986**, *42*, 1621–1631.

21. L.T. Scott, M.M. Boorum, B.J. McMahon, S. Hagen, J. Mack, J. Blank, H. Wegner and A. de Meijere, A rational chemical synthesis of C-60. *Science* **2002**, *295*, 1500–1503.

22. R.B. Woodward, T. Fukunaga and R.C. Kelly, Triquinacene. *J. Am. Chem. Soc.* **1964**, *86*, 3162–3164.

23. I.T. Jacobson, Polyquinanes. A simple synthesis of tricyclo[5.2.1.0^{4,10}]deca2,5,8-triene. *Acta Chem. Scand.* **1967**, *21*, 2235–2246.

24. P.E. Eaton, W.H. Bunnelle and P. Engel, Synthesis of dodecahedrane precursors. 3. Synthesis of alkylidene-1,3-cyclopentanediones and attempts to roof peristylanes. *Can. J. Chem.* **1984**, *62*, 2612–2626.

25. R.L. Sobczak, M.E. Osborn and L.A. Paquette, Functionalized (C_S)-C17-heptaquinone derivatives. Chemical transformations along the fluted perimeter of a topologically spherical molecule. *J. Org. Chem.* **1979**, *44*, 4886–4890.

26. M.J. Wyvratt and L.A. Paquette, Domino Diels-Alder reactions II. A four-step conversion of cyclopentadiene to triquinacene. *Tetrahedron Lett.* **1974**, *41*, 2433–2436.

27. P.W. Codding, K.A. Kerr, A. Oudeman and T.S. Sorensen, Tricarbonyl (triquinacene)-molybdenum and -tungsten. *J. Organomet. Chem.* **1982**, *232*, 193–199.

28. L.A. Paquette, Dodecahedranes and allied spherical molecules. *Chem. Rev.* **1989**, *89*, 1051–1065.

29. L.A. Paquette, D.W. Balogh, R. Usha, D. Kountz and G.G. Christoph, Crystal and molecular structure of a pentagonal dodecahedrane. *Science* **1981**, *211*, 575–576.

30. H. Prinzbach and K. Weber, From an insecticide to Plato's universe — the pagodane route to dodecahedranes: new pathways and new perspectives. *Angew. Chem. Int. Ed. Engl.* **1994**, *33*, 2239–2257.

31. A.P. Marchand and A.D. Earlywine, Heptacyclo[5.5.1.14,10^{02},6^{03},11^{05}.9^{08},12]-tetradecane-13,14-dione: a novel, polycyclic perpendobiplanar D_{2d} diketone. *J. Org. Chem.* **1984**, *49*, 1660–1661.

32. L.A. Paquette, A.R. Browne, C.W. Doecke and R.V. Williams, Short, stereocontrolled synthesis of [4]peristylane. *J. Am. Chem. Soc.* **1983**, *105*, 4113–4115.

33. C.-H. Lee, S. Liang, T. Haumann, R. Boese and A. de Meijere, *p*-[3^2.5^6] Octahedrane, the $(CH)_{12}$ hydrocarbon with D_{3d} symmetry. *Angew. Chem. Int. Ed. Engl.* **1993**, *32*, 559–561.

34. H. Okamoto, K. Satake, H. Ishida and M. Kimura, Photoreaction of a 2,11-diaza[3.3]paracyclophane derivative; formation of an octahedrane by photochemical dimerization of benzene. *J. Am. Chem. Soc.* **2006**, *128*, 16508–16509.

35. H.-D. Martin and P. Pföhler, Pentacyclo[6.4.0.0^{2,5}.0^{3,10}.0^{4,9}]dodeca-6,11-diene, an *o,o':o,p'* dimer of benzene. *Angew. Chem. Int. Ed. Engl.* **1978**, *11*, 847–848.

36. P.E. Eaton, A. Srikrishna and F. Uggieri, Cyclopentannulation of bicyclo[3.3.0] octane-3,7-dione. A more convenient synthesis of the [5]peristylane system. *J. Org. Chem.* **1984**, *49*, 1728–1732.

37. C.-C. Shen and L.A. Paquette, Development of a strategy for the synthesis of the spherical hydrocarbon *p*-[4^2.5^8]decahedrane. *Tetrahedron* **1994**, *50*, 4949–4956.

38. H.W. Kroto, J.R. Heath, S.C. O'Brien, R.F. Curl and R.E. Smalley, C_{60}: Buckministerfullerene. *Nature* **1985**, *318*, 162–163.

39. S. Liu, Y-J. Lu, M.M. Kappes and J.A. Ibers, The structure of the C_{60} molecule — X-ray crystal structure determination of a twin at 110 K. *Science* **1991**, *254*, 408–410.

40. W. Krätschmer, L.D. Lamb, K. Fostirolopolous and D.R. Huffman, Solid C_{60}: a new form of carbon. *Nature* **1990**, *347*, 354–358.

41. H.K. Gupta, P.E. Lock and M.J. McGlinchey, Metal complexes of trindane: possible precursors of sumanene. *Organometallics* **1997**, *16*, 3628–3634.

42. H.K. Gupta, P.E. Lock, D.W. Hughes and M.J. McGlinchey, Trindane-ruthenium sandwiches: an NMR and X-ray crystallographic study of [(trindane)RuCl$_2$]$_2$, (trindane)RuCl$_2$[P(OMe)$_3$], and [(trindane)$_2$Ru][BF$_4$]. *Organometallics* **1997**, *16*, 4355–4361.

43. T.L. Tisch, T.J. Lynch and R. Dominguez, Polymetallic compounds of extended polyaromatic ligands. Rhenium and manganese carbonyl derivatives of truxene. *J. Organomet. Chem.* **1989**, *377*, 265–273.

44. P.W. Rabideau and A. Sygula, Buckybowls: polynuclear aromatic hydrocarbons related to the buckminsterfullerene surface. *Acc. Chem. Res.* **1996**, *29*, 235–242.

45. A. Sygula and P.W. Rabideau, Non-pyrolytic syntheses of buckyballs: corannulene, cyclopentacorannulene, and a semibuckminsterfullerene. *J. Am. Chem. Soc.* **1999**, *121*, 7800–7803.

46. M. Schmittel and C. Vavilala, Kinetic isotope effects in the thermal C^2-C^6 cyclization of enyne-allenes: experimental evidence supports a stepwise mechanism. *J. Org. Chem.* **2005**, *70*, 4865–4868.

47. W.E. Barth and R.G. Lawton, Dibenzo[*ghi,mno*]fluoranthene. *J. Am. Chem. Soc.* **1966**, *88*, 380–381.

48. L.T. Scott, M.M. Hashemi, D.T. Meyer and H.B. Warren, Corannulene. A convenient new synthesis. *J. Am. Chem. Soc.* **1991**, *113*, 1082–1084.

49. T.J. Seiders, E.L. Elliott, G.H. Grube and J.S. Siegel, Synthesis of corannulene and alkyl derivatives of corannulene. *J. Am. Chem. Soc.* **1999**, *121*, 7804–7813.

50. A.M. Butterfield, B. Gilomen and J.S. Siegel, Kilogram-scale production of corannulene. *Org. Process Res. Dev.* **2012**, *16*, 664–676.

51. Y.T. Yu and J.S. Siegel, Aromatic molecular-bowl hydrocarbons: synthetic derivatives, their structures and physical properties. *Chem. Rev.* **2006**, *106*, 4843–4867.

52. G. Mehta, S.R. Shah and K. Ravikumar, Towards the design of tricyclo[*def*, *jkl,pqr*]triphenylene (Sumanene): a 'bowl-shaped' hydrocarbon featuring a structural motif present in C_{60} (Buckminsterfullerene). *J. Chem. Soc. Chem. Commun.* **1993**, 1006–1008.

53. H. Sakurai, T. Daiko and T. Hirao, A synthesis of sumanene, a fullerene fragment. *Science* **2003**, *301*, 1878.

54. T. Amaya and T. Hirao, Chemistry of sumanene. *Chem. Rec.* **2015**, *15*, 310–321.

55. S. Alvi and R. Ali, Synthetic approaches to bowl-shaped π-conjugated sumanene and its congeners. *Beilstein J. Org. Chem.* **2020**, *16*, 2212–2259.

56. A. Sekiguchi, T. Yatabe, C. Kabuto and H. Sakurai, Chemistry of organosilicon compounds. 303. The missing hexasilaprismane: synthesis, x-ray analysis and photochemical reactions. *J. Am. Chem. Soc.* **1993**, *115*, 5853–5854.

57. A. Sekiguchi, C. Kabuto and H. Sakurai, [(Me_3Si)$_2$CHGe]$_6$, the first hexagermaprismane. *Angew. Chem. Int. Ed.* **1989**, *101*, 97–98.

58. R.C. Burns, R.J. Gillespie, W.-C. Luk and D.R. Slim, Preparation, spectroscopic properties, and crystal structures of $Te_6(AsF_6)_4$.$2AsF_3$ and $Te_6(AsF_6)_4$.$2SO_2$: a new trigonal-prismantic cluster cation, hexatellurium(4+) *Inorg. Chem.* **1979**, *18*, 3086–3094.

59. R.C. Burns, R.J. Gillespie, J.A. Barnes and M.J. McGlinchey, Molecular orbital investigation of the structure of some polyatomic cations and anions of the main-group elements. *Inorg. Chem.* **1982**, *21*, 799–807.

60. H. Matsumoto, K. Higuchi, Y. Hoshino, H. Koike, Y. Naoi and Y. Nagai, The first octasilacubane system: synthesis of octakis-(t-butyldimethylsilylpentacyclo[4.2.0.0^{2,5}.0^{3,8}.0^{4,7}]octasilane. *J. Chem. Soc. Chem. Chem. Commun.* **1988**, 1083–1084.

61. A. Sekiguchi, T. Yatabe, H. Kamatani, C. Kabuto and H. Sakurai, Chemistry of organosilicon compounds. 293. Preparation, characterization, and crystal structures of octasilacubanes and octagermacubanes. *J. Am. Chem. Soc.* **1992**, *114*, 6260–6262.

62. K. Furukawa, M. Fujino and N. Matsumoto, Cubic silicon cluster. *Appl. Phys. Lett.* **1992**, *60*, 2744–2745.

63. L.R. Sita and L. Kinoshita, Octakis(2,6-diethylphenyl)octastannacubane. *Organometallics* **1990**, *9*, 2865–2867.

64. L.R. Sita and R.D. Bickerstaff, 2,2,4,4,5,5-hexakis(2,6-diethylphenyl) pentastanna[1.1.1]propellane: characterization and molecular structure. *J. Am. Chem. Soc.* **1989**, *111*, 6454–6456.

65. L.R. Sita and I. Kinoshita, Decakis(2,6-diethylphenyl)decastanna[5]prismane: characterization and molecular structure. *J. Am. Chem. Soc.* **1991**, *113*, 1856–1857.

66. A. Sekiguchi and H. Sakurai, Cage and cluster compounds of silicon, germanium and tin. In *Advances in Organometallic Chemistry*, Academic Press: New York, USA, 1995; Vol. 37, pp. 1–38.

67. C. Liu and Z-M. Sun, Recent advances in structural chemistry of Group 14 Zintl ions. *Coord. Chem. Rev.* **2019**, *382*, 32–56.

68. C.E. Housecroft, *Cluster Molecules of the p-Block Elements*, Oxford University Press: Oxford, UK, 1994; pp. 45–51.

69. D.M.P. Mingos and D.J. Wales, *Introduction to Cluster Chemistry*, Prentice Hall, Englewood Cliffs: New Jersey, USA, 1990; pp. 173–177.

70. D. Tian, J. Zhao, B. Wang and R.B. King, Dual relationship between large gold clusters (antifullerenes) and carbon fullerenes: a new lowest-energy cage structure for Au_{50}. *J. Phys. Chem. A* **2007**, *111*, 411–414.

71. W.N. Lipscomb and L. Massa, Examples of large closo boron hydride analogs of carbon fullerenes. *Inorg. Chem.* **1992**, *31*, 2297–2299.

72. A. Derecskei-Kovacs, B.L. Dunlap, W.N. Lipscomb, A. Lowrey, D.S. Marynick and L. Massa, Quantum chemical studies of boron fullerene analogs. *Inorg. Chem.* **1994**, *33*, 5617–5619.

73. A. Gindulyte, W.N. Lipscomb and L. Massa, Proposed boron nanotubes. *Inorg. Chem.* **1998**, *37*, 6544–6548.

74. J. Bicerano, D.S. Marynick and W.N. Lipscomb, Molecular orbital studies on large closo boron hydrides. *Inorg. Chem.* **1978**, *17*, 3443–3453.

Chapter 8
Enhancing Rotational Symmetry

"The law of right-left symmetry was used in classical physics but was not of any great practical importance there. One reason for this derives from the fact that right-left symmetry is a discrete symmetry, unlike rotational symmetry, which is continuous."

— *Chen-Ning Yang*

Rotational symmetry has always attracted artists and architects, landscape gardeners and scientists alike, whether it be a rose window in a cathedral, or a perfect flower (Figure 8.1), an Abelian group or a cyclic molecule. Perhaps the most famous, probably apocryphal, example in chemistry is Kekulé's dream of a snake biting its tail that prompted him to suggest a cyclic structure for benzene. Since then, the syntheses of rings of all sizes, small or large, simple or highly decorated, have continued to attract great attention. We here discuss how imaginative scientists have tackled these synthetic problems, and where the current challenges lie.

8.1. Chauvin's Carbomers

Any discussion of enhanced rotational symmetry must encompass the major contributions of Remi Chauvin and his group at Université Paul Sabatier in Toulouse, a beautiful French city known as *La Ville*

Figure 8.1. Rotational symmetry in a rose window and in a sunflower.

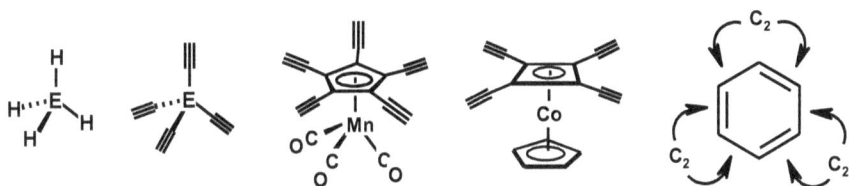

Figure 8.2. Insertion of C_2 units into EH_4, into the C–H bonds in cyclopentadienyl or cyclobutadiene rings in organometallic complexes, and into the ring bonds of benzene.

Rose in recognition of the pink colour of so many of its buildings. In 1995, he introduced the idea of *carbomers* in which (i) a –C≡C– unit is inserted into each single bond in a molecule, or (ii) a =C=C= unit is inserted into each double bond in a molecule, or (iii) a ≡C–C≡ unit is inserted into each triple bond in a molecule.[1] These operations leave the symmetry of the molecule unchanged, and in case (ii) the orthogonal nature of the π bonds in the allene linkage guarantees that the planar substitution pattern is maintained.

Among the numerous examples that were suggested, one can cite the known molecules E(C≡CH)$_4$, where E = C, Si, Ge or Sn,[2,3] as carbomers of EH_4, or the peralkynylated half-sandwich organometallic complexes shown in Figure 8.2.[4] More intriguingly, Chauvin discussed carbomers of benzene, firstly, in which a C_2 unit has been inserted into each of the ring bonds, as in $C_{18}H_6$, **1** (which would give rise to two resonance forms of dodecadehydro[18]annulene, Figure 8.3), and

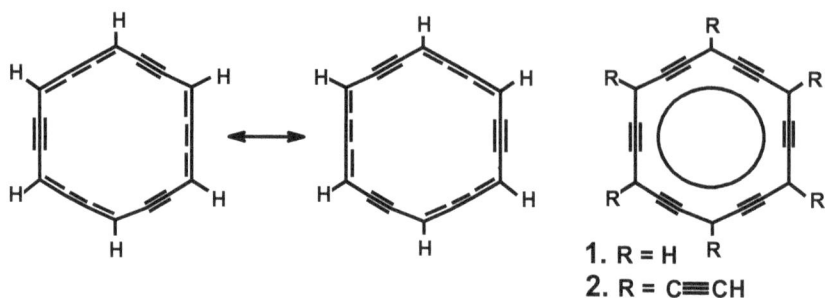

Figure 8.3. Resonance forms of *carbo*-benzene, **1**.

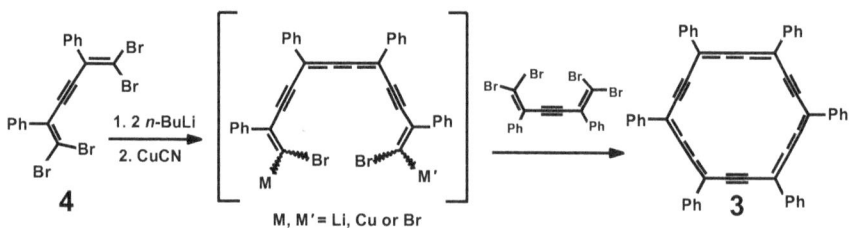

Scheme 8.1. Synthesis of hexaphenyl-*carbo*-benzene, **3**.

likewise into hexaethynylbenzene to form **2**, that is designated as the "total carbomer" of benzene. High-level calculations suggest that *carbo*-benzene sustains a strong diatropic ring current and has aromatic character comparable to benzene itself.

Although the parent compound, **1**, is still not available, hexa-aryl substituted versions, such as $C_{18}Ph_6$, **3**, have been prepared by copper-mediated coupling of three diphenyltetrabromohexa-1,5-dien-3-yne units, **4**; the culminating steps are shown in Scheme 8.1, and X-ray crystallographic data revealed a planar D_{6h} structure.[5,6] Impressively, **5**, the carbomer of an octaarylnaphthalene, although available only in low yield, has also been characterised by X-ray crystallography (Figure 8.4).[7]

In related work, Scott has reported the syntheses of a series of pericyclynes that could be regarded as carbomers of the cycloalkanes ranging from cyclopentane to cyclooctane in which a C_2 unit has been inserted into each carbon-carbon ring bond (Figure 8.5).[8]

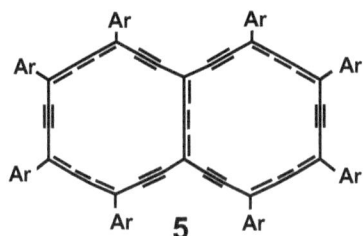

Figure 8.4. A *carbo*-octaarylnaphthalene, **5**, Ar = *para-n*-pentylphenyl.

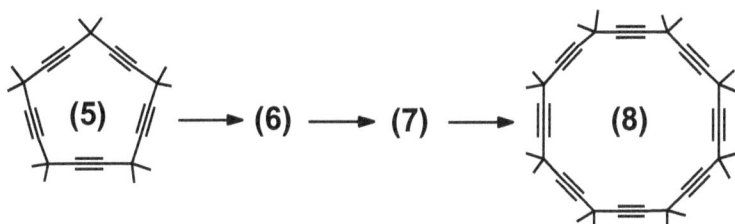

Figure 8.5. A selection of Scott's pericyclynes.

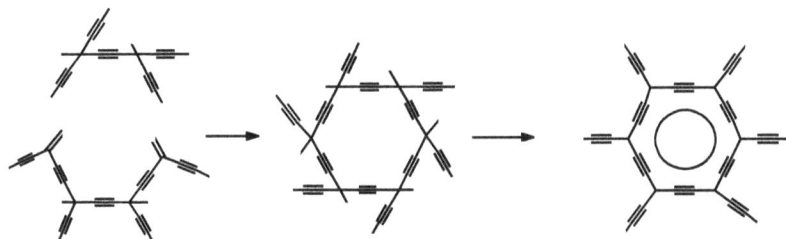

Scheme 8.2. Synthesis of the silylated "total carbomer" of benzene, **7**. Reagents: (a) Co$_2$(CO)$_8$, (b) SnCl$_2$/HCl, (c) ceric ammonium nitrate.

This concept has been elegantly extended by Chauvin to form a functionalised [6]pericyclyne, **6**, bearing six alkyne and six methoxy substituents. The final step of the synthesis is shown in Scheme 8.2. After desilylation, the resulting X-ray crystal structural characterisation (Figure 8.6) revealed that the molecule adopted a chair conformation, typical of a D_{3d}-symmetric cyclitol, but with greatly extended

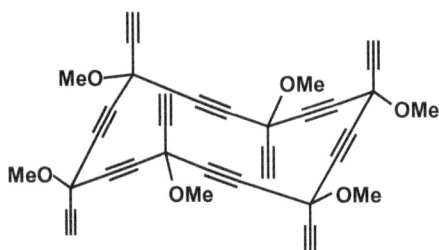

Figure 8.6.   Chair conformation of a hexaalkynyl-hexamethoxy[6]pericyclyne.

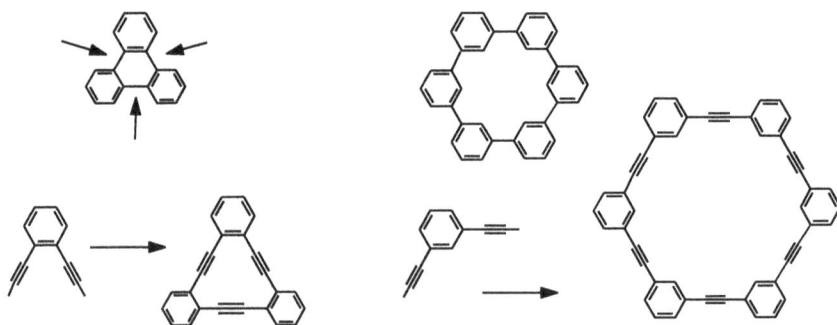

Scheme 8.3.   Alkyne metathesis of *ortho-* and *meta*-diethynylarenes to form pericyclynes.

linear sections (~4.144 Å) between the sp^3 carbons.[9] Cobalt-assisted aromatisation of **6** delivered molecule **7**, in which C_2 units have been inserted not only into the benzene ring linkages, but also into the external bonds. This represents a silylated version of **2**, the "total carbomer" of benzene.

In other work, **8** and **9**, the three-fold and six-fold symmetric carbomers of triphenylene and hexa-*m*-phenylene,[10] respectively, have been prepared by alkyne metathesis of an *ortho-* or a *meta*-diethynyl-benzene (Scheme 8.3).[11] Moreover, the syntheses of a wide range of heterocyclic pericyclynes, containing such elements as silicon, germanium or phosphorus, have been reported (Figure 8.7).[12,13]

Also particularly worthy of mention is Diederich's cubane carbomer, **10**,[14] whereby a diyne unit has been incorporated into each of the cube's 12 edges, thus retaining the O_h symmetry of the parent

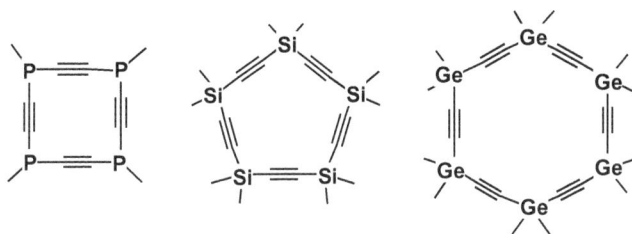

Figure 8.7. Selected heterocyclic pericyclynes.

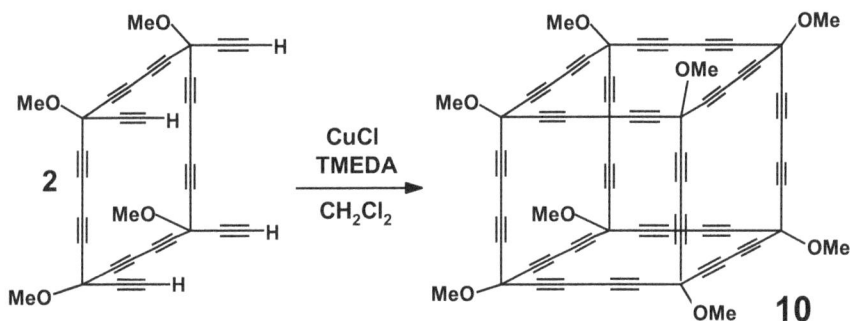

Scheme 8.4. Diederich's route to the polyalkynylated cubane, **10**.

hydrocarbon. The final stage of the synthesis involves copper-mediated coupling of two "half-cube" fragments (Scheme 8.4).

8.2. Fitjer's Universal Rotane Synthesis

The extension of rotational symmetry out of the plane is exemplified by [3]rotane in which each vertex of the central cyclopropyl ring is spiro-linked to another cyclopropyl moiety. Among the routes that have been adopted, the one illustrated in Scheme 8.5 is typical. Starting from bicyclopropylidene, **11**, addition of chloromethylcarbene to form **12** was followed by elimination of HCl to generate the spiro alkene, **13**, and finally Simmons-Smith cyclopropanation to yield the desired product, **14**.[15] The X-ray crystal structure of [3]rotane has been determined and revealed the expected D_{3h} symmetry, in which the bonds within and directly linked to the central ring are noticeably

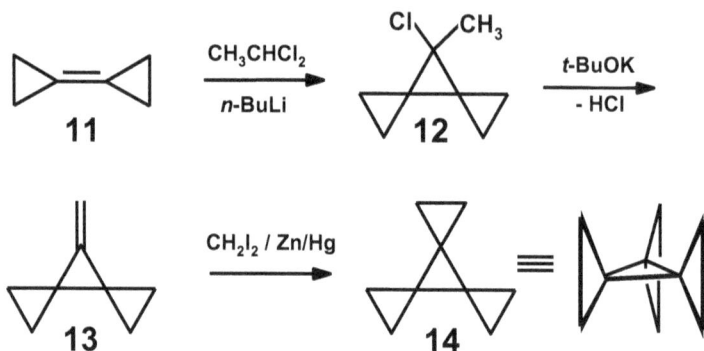

Scheme 8.5. A synthetic route to [3]rotane.

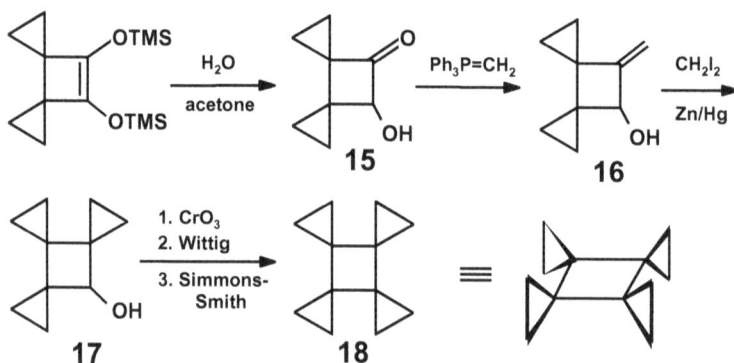

Scheme 8.6. A synthetic route to [4.3]rotane.

shorter (1.476 Å) than those forming the external CH_2–CH_2 fragments in the peripheral rings (1.525 Å).[16]

Other approaches have been used to prepare a variety of different rotanes such as [4.3]rotane. (In this nomenclatural system, the first number indicates size of the central ring, and the second the ring size of the peripheral substituents.) As shown in Scheme 8.6, the dicyclopropyl-acyloin, **15**, was treated with a Wittig reagent to introduce a methylidene unit in **16**, that was subsequently cyclopropanated to form **17**. Oxidation to the ketone, and then another Wittig reaction was followed by a final Simmons-Smith cyclopropanation to furnish **18**.[17]

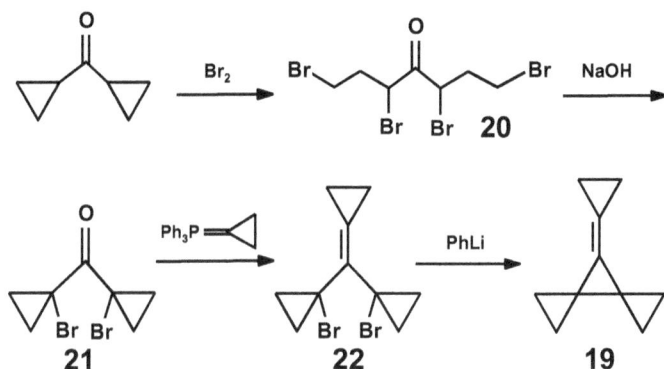

Scheme 8.7. Preparation of cyclopropylidenedispiroheptane.

However, these methods did not always provide sufficient material for further studies, and a more general synthetic route was clearly desirable. Gratifyingly, in a brilliant new approach, Lutz Fitjer from the University of Göttingen in Germany developed what has been termed a "Universal Rotane Synthesis", by which means rotanes could be continually sequentially expanded to the next in the series (cyclopropylogation). The crucial precursor for this procedure is the cyclopropylidene derivative, **19**, prepared as shown in Scheme 8.7. Bromination of dicyclopropyl ketone opens both rings as in **20**, and reaction with base results in double dehydrobromination to form 1,1'-dibromodicyclopropyl ketone, **21**, that can be converted into the cyclopropylidene derivative, **22**, by using the appropriate Wittig reagent; finally, ring closing with phenyl lithium yields cyclopropylidenedispiroheptane, **19**.[18]

Now, starting from **19**, cycloaddition with *p*-nitrobenzenesulfonyl azide generates the transient triazole, **23**, that suffers loss of dinitrogen and undergoes ring expansion to form the imine, **24**, and hydrolysis releases the ring-expanded ketone, **25** (Scheme 8.8). At this point, one can now choose whether to form [4.3]rotane by reaction of **25** with $Ph_3P=CH_2$ and then Simmons-Smith cyclopropanation, or to react with the cyclopropylidene Wittig reagent to generate the precursor **26** poised for the next expansion to [5.3]rotane, and beyond. Multiple repeats of this sequence allow straightforward formation of a succession of [*n*.3]rotanes.[19]

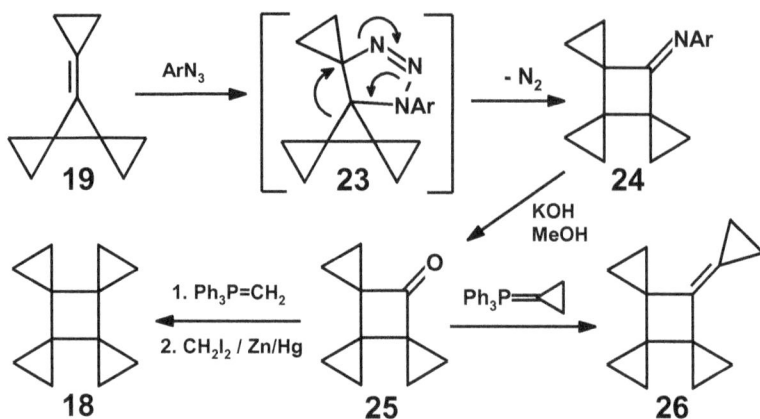

Scheme 8.8. Fitjer's universal rotane synthesis, where $ArN_3 = p\text{-}O_2N\text{-}C_6H_4\text{-}SO_2N_3$.

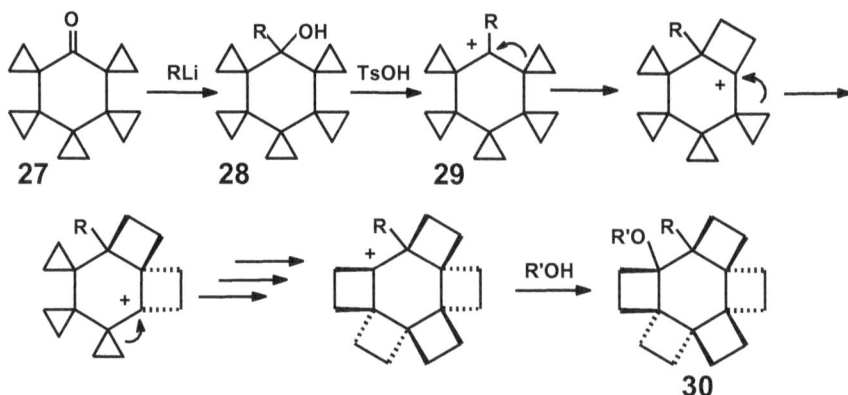

Scheme 8.9. A sequence of carbocationic rearrangements converting five spiro-cyclopropyl substituents into fused cyclobutyl rings.

However, these observations are merely the start of another dramatic sequence of ring expansions discovered by Fitjer. Reduction of the ketone **27** to alcohol **28**, followed by protonation to form the cation **29** prompts a spectacular series of ring expansions whereby each cyclopropyl substituent is sequentially transformed into a cyclobutyl fragment fused to the central ring, resulting finally in **30** (Scheme 8.9).[20]

This is a general phenomenon as exemplified by the cascade rearrangement of the cyclohexanone **31** bearing five spiro-bonded

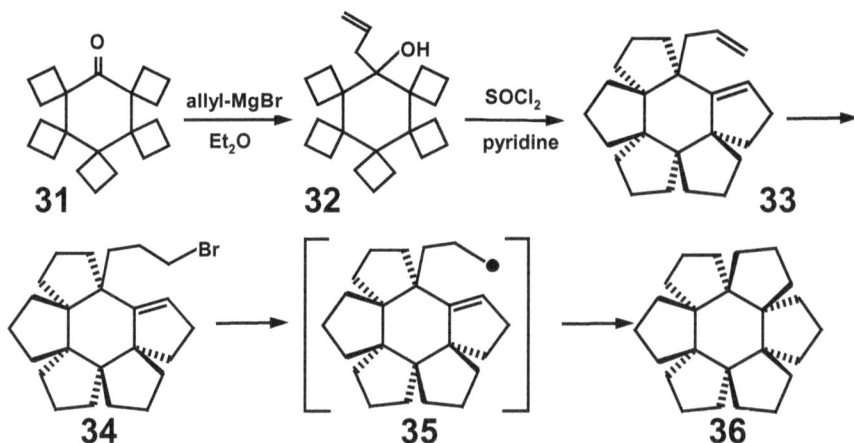

Scheme 8.10. A sequence of carbocationic rearrangements converting five spiro-cyclobutyl substituents into fused cyclopentyl rings, with final ring closure to form [6.5]coronane, **36**.

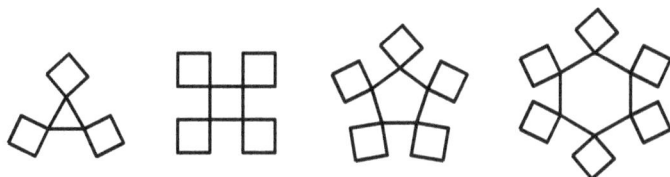

Figure 8.8. [3.4]rotane, [4.4]rotane, [5.4]rotane and [6.4]rotane.

cyclobutyl substituents (Scheme 8.10).[21] In this case, prior incorporation of a allyl group, as in **32**, leads eventually, after bromination of **33**, and tributyltin-mediated ring closure of **34**, via the intermediate radical **35**, to [6.5]coronane, **36**.

Finally, we note that in an impressively comprehensive report, Fitjer described the syntheses and X-ray crystal structures of a complete series of rotanes in which cyclobutyl groups are spiro-linked to 3-, 4-, 5- and 6-membered central rings (Figure 8.8).[22]

It is noteworthy that multiple insertion of C_2 units into [n.3] rotanes have led to the formation of very large rings, denoted by de Meijere as "exploded" pericyclynes (Figure 8.9). These include cases where $n = 1$ or 2 leading to 25- and 30-membered rings. An extreme example, $n = 8$, is seen in **37** in which the incorporation of a dialkyne

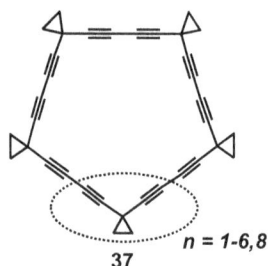

Figure 8.9. Examples of "exploded" pericyclynes.

moiety between each pair of cyclopropyls in [12]rotane results in formation of a C_{60} ring adopting an enormously elongated chair conformation of D_{3d} symmetry that has been characterised spectroscopically and by X-ray crystallography.[23,24]

We close this section on cyclopropanation with de Meijere's characterisation of tetracyclopropylmethane, **38**. Although the tetracyclopropyl derivatives $(C_3H_5)_4E$, where E is silicon, germanium or tin, were already known,[25] the steric problems arising in the carbon case are severe. Treatment of dicyclopropyldiethenylmethane, **39**, under Simmons-Smith conditions was unsuccessful, and reaction with diazomethane catalysed by $Pd(OAc)_2$ gave **38** in only 13% yield. However, multiple repeats of this procedure eventually furnished the product in 92% yield.[26] The X-ray crystal structure of tetracyclopropylmethane revealed that the molecule does not adopt a D_{2d} structure, as do the Si, Ge or Sn analogues, or in $C(CH_2O_2CCl_3)_4$ where the trichloroacetate groups are arranged pairwise,[27] but instead the bulky rings in **38** cause a twisting such that the molecule has S_4 symmetry (Scheme 8.11).

8.3. Functionalised Polycyclics of 3- or 5-fold Symmetry

The structures of sumanene and corannulene, the widely studied fragments of C_{60}, offer tempting opportunities to enhance their rotational symmetries, and numerous aesthetically pleasing molecules have been reported. Typically, the three-fold "flower petal" arrangement of sumanene has been augmented by adding organic functional groups to the periphery, as in **40**, and also by the preparation of

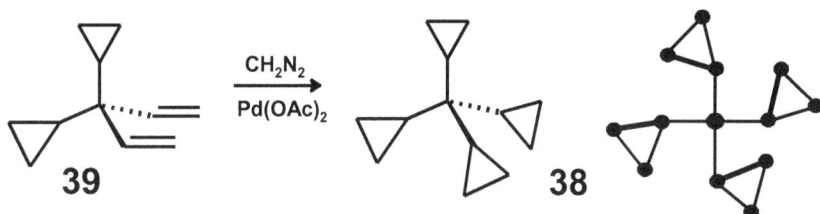

Scheme 8.11. Preparation of tetracyclopropylmethane, **38**.

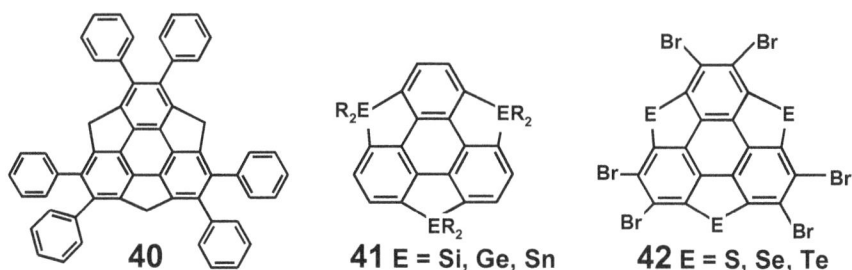

Figure 8.10. Functionalised sumanenes of enhanced three-fold symmetry.

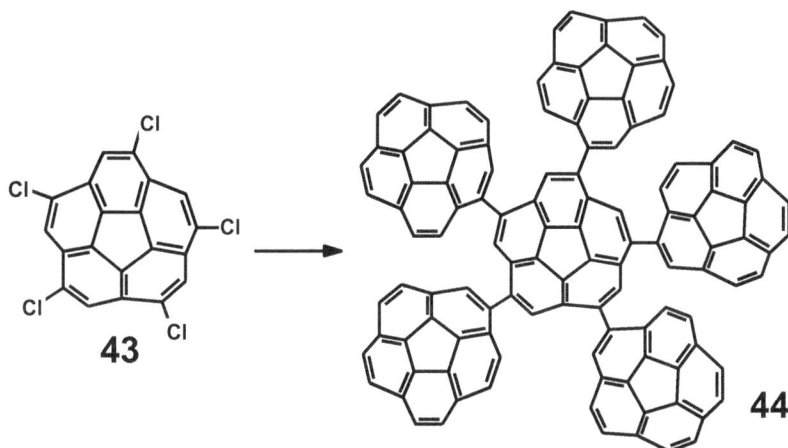

Scheme 8.12. Five-fold functionalisation of corannulene.

hetero-sumanenes containing a very wide variety of main group elements, exemplified by **41** and **42** (Figure 8.10).[28,29]

Likewise, the corannulene system has been multiply substituted yet still maintained its five-fold symmetry. As examples, one can cite

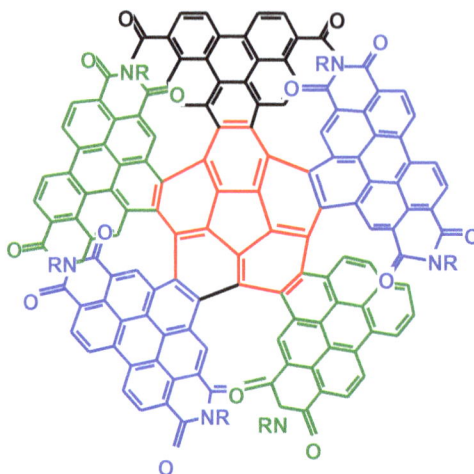

Figure 8.11. Corannuranylene pentapetalae, **45**.

the formation of the 1,3,5,7,9-pentachloro derivative, **43**, by treatment of corannulene with iodine monochloride, *en route* to 1,3,5,7,9-pentacorannulenylcorannulene, **44** (Scheme 8.12).[30]

However, the crème de la crème must surely be the *corannurylene pentapetalae*, **45**, reported by Li, Siegel and Wang, and shown in Figure 8.11.[31] In this system, the five peripheral six-membered rings have each been decorated by addition of a large polyaromatic moiety thus generating spectacularly picturesque non-planar graphenoids. Interestingly, the wingspans of the external fragments are such that considerable overlap occurs, thus giving rise to pentapetalae of differing molecular symmetry. When each overlaps uniformly in turn with its next neighbour, the resulting structure is D_5-symmetric. However, other conformations are possible, and both D_5 and C_2 structures have been characterised by X-ray crystallography. Moreover, it was also reported that transistor devices demonstrated that, without any π-π stacking, the honeycomb lattice could also support electron transport.[31]

8.4. The Quest for Perfluoroferrocene

8.4.1. *Persubstituted Ring Systems*

While the preparation of rotationally symmetric complex systems of increasing radial size, such as the pericyclyne **37**, or the heavily decorated corannulenes **44** and **45**, has certainly been impressive, the search for apparently much simpler molecules also continues to attract attention. For example, molecular fragments of the type $[Ph_nC_n]^{n\pm}$ exist either as neutral or charged species, or as components of organometallic systems, such as $(C_3Ph_3)Ni(C_5H_5)$, (C_4Ph_4) $Co(C_5H_5)$ and $(C_5Ph_5)_2Fe$. They have been investigated for their relevance to the Hückel "4n + 2 rule" and the phenomenon of aromaticity.

A feature of these D_n-symmetric propeller-type systems is the dihedral angle, θ, made by the peripheral phenyls with the central ring. This represents a compromise since the coplanar arrangement, θ = 0°, that maximises π overlap may induce serious steric strain, whereas in the orthogonal conformer, θ = 90°, although interactions between bulky groups are minimised, π conjugation is disrupted. For the series C_nPh_n (n = 3–7), the angle subtended at the centre of the internal ring by adjacent phenyls (ω = 360°$/n$) decreases in value from 120°, 90°, 72°, 60°, to 51.4°, respectively. Although increasing the ring size lengthens the radial distance of the external phenyls from the ring centre, this is more than compensated for by the decreasing value of ω, thus placing the external phenyls in a more restricted locale. In the triphenylcyclopropyl cation, the value of θ is approximately 5°, in tetraphenylcyclobutadiene-metal complexes it averages about 35°, in pentaphenylcyclopentadienyl cases θ is ca. 50°, and in C_6Ph_6 75°.[32] In the event, when the structure of the heptaphenyltropylium cation, **46**, was eventually determined,[33] almost 40 years after its initial synthesis,[34] the average dihedral angle θ was found to be 80°, but more significantly, the molecule was no longer planar and instead adopted a shallow boat conformation.[33] It is noteworthy that the corresponding heptamethyltropylium cation, **47**, is also a boat, and although it can form metal complexes, as in $(\eta^5\text{-}C_7Me_7)W(CO)_3I$, **48**, only five of the ring carbons participate in bonding to the metal (Scheme 8.13).[35]

8.4.2. *Polyhalogenated Metal Sandwich Complexes*

Since the discovery of ferrocene more than 50 years ago, the chemistry of sandwich compounds has continued to grow exponentially.

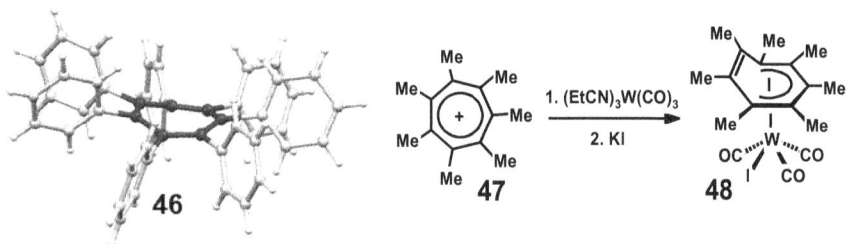

Scheme 8.13. Molecular structure of $[C_7Ph_7]^+$, **46**, and of $(C_7Me_7)W(CO)_3I$, **48**.

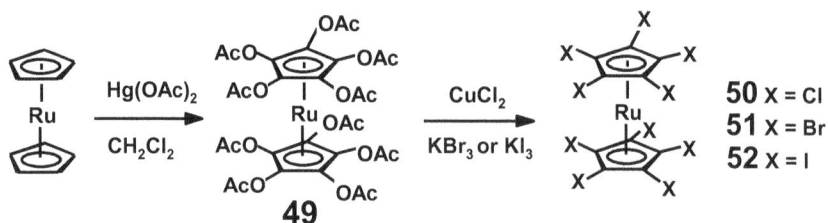

Scheme 8.14. Preparation of perhalogenated ruthenocenes.

Typically, fully alkylated or arylated molecules such as decamethyl- and decaphenyl-ferrocene have long been known,[36,37] and ferroceno-phanes bearing five bridging links have also been synthesised.[38] Moreover, fully halogenated sandwich compounds $(C_5X_5)_2M$, where M = Fe or Ru, and X = Cl, Br, I, are known, but the corresponding fluoro derivatives are still not available.

The first report of the preparation of decachloroferrocene by Hedberg and Rosenberg[39] in 1973 involved repeated lithiation of ferrocene and then reaction with hexachloroethane. However, this procedure also generated a mixture of chloroferrocenes that necessi-tated extensive chromatographic separations. Twenty years later, Winter and Rheingold described the decamercuration of ruthenocene with mercuric acetate in CH_2Cl_2 to form **49** in 88% yield. Subsequent reaction with $CuCl_2$, KBr_3 or KI_3 in refluxing acetone yielded deca-chloro-, decabromo- or decaiodoruthenocene, **50**–**52** (Scheme 8.14); $(C_5Br_5)_2Ru$ was characterised by X-ray crystallography.[40] Convenient syntheses of $(C_5X_5)Fe(C_5H_5)$, where X = Cl or Br, have also been reported.[41,42]

8.4.3. *Metal Complexes of Hexafluorobenzene*

Considering now ring systems of the type C_nF_n, we are unaware of any reported chemistry of the 1,2,3-trifluorocyclopropenyl group, but the next congener tetrafluorocyclobutadiene, C_4F_4, was successfully prepared by David Lemal from Dartmouth College in New Hampshire, an inspiring leader in this field. The synthesis was accomplished by ozonolysis of Dewar-hexafluorobenzene, **53**, to form the strongly acidic tetrafluorocyclobutene-3,4-dicarboxylic acid, **54**, followed by dehydration with phosphorus pentoxide to yield the anhydride, **55**. Subsequent photolysis at 2537 Å in the presence of nitrogen brought about elimination of CO and CO_2 to form the short-lived tetrafluorocyclobutadiene, **56**, that rapidly dimerised to generate octafluoro[4.2.0.0^{2,5}]cycloocta-3,7-diene, **57** (Scheme 8.15). In the absence of nitrogen, this photolysis experiment yielded its isomer octafluorocyclooctatetraene, **58**. Even more convincing proof of the intermediacy of tetrafluorocyclobutadiene was the photolysis of **55** in the presence of furan that delivered the Diels-Alder cycloadduct **59**.[43]

We defer treatment of C_5F_5 to the next section in order to discuss briefly the organometallic chemistry of C_6F_6, which is conveniently available commercially. As noted above, hexafluorobenzene can exist not only in the normal hexagonal form, but also as its Dewar isomer, **53**, prepared by photolysis. This latter molecule reacts with metal carbonyl anions, such as $[Re(CO)_5]^-$ or $[(C_5H_5)Fe(CO)_2]^-$, with displacement of a vinylic fluorine (Scheme 8.16).[44]

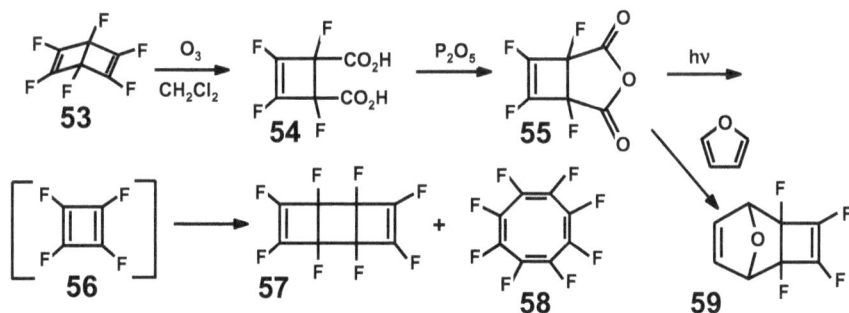

Scheme 8.15. Preparation and reactions of tetrafluorobutadiene.

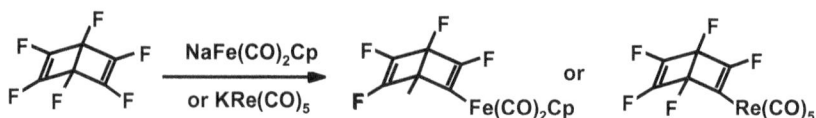

Scheme 8.16. Reactions of metal carbonyl anions with Dewar-hexafluorobenzene.

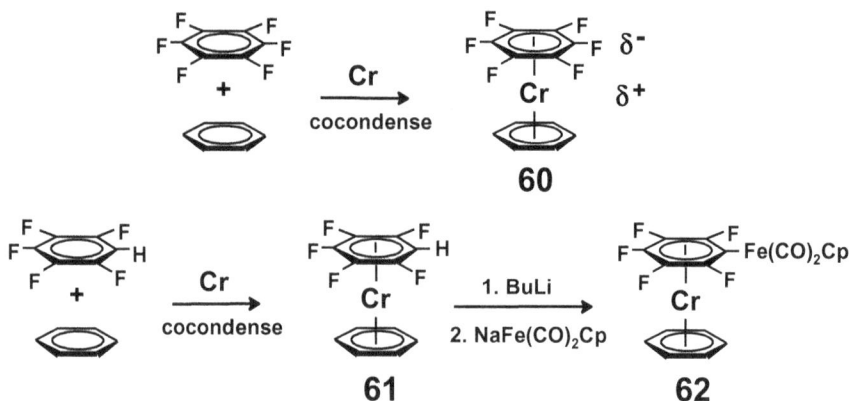

Scheme 8.17. Syntheses of $(\eta^6\text{-}C_6H_6)Cr(\eta^6\text{-}C_6F_5X)$ complexes.

However, π-bonded fluoroarene-metal complexes have also been prepared, primarily by the metal atom co-condensation procedure. This technique, pioneered by Peter Timms at the University of Bristol, involves the thermal evaporation of a metal, under vacuum, onto the liquid nitrogen-cooled walls of a container with concomitant deposition of an arene, phosphine or other volatile ligand.[45-47] Typically, co-condensation of chlorobenzene and chromium provides a convenient route to bis(chlorobenzene)chromium(0), $(C_6H_5Cl)_2Cr$,[48] that is not preparable by the normal Fischer-Hafner method[49] since the chloro substituent reacts with the added $AlCl_3$. This new approach (Scheme 8.17) allowed the first preparation of a π-complexed hexafluorobenzene complex by co-condensation of hexafluorobenzene, benzene and chromium to form $(\eta^6\text{-}C_6F_6)Cr(\eta^6\text{-}C_6H_6)$, **60**, which exhibits well-resolved 1:6:15:20:15:6:1 septets in both the ^1H and ^{19}F NMR regimes.[50,51]

We note that a fascinating and eminently readable historical perspective[52] on the discovery of the bis(arene)chromium complexes by

Ernst Otto Fischer and Walter Hafner has been written by Dietmar Seyferth, the founding editor of *Organometallics*. It is particularly noteworthy that, although bis(arene)chromium complexes are normally very susceptible to aerial oxidation at the chromium centre, the novel sandwich compound **60** is remarkably air-stable for months. This may be ascribed to an "internal oxidation" involving partial electron transfer from the metal to the fluorinated ring, as represented in Scheme 8.17. Support for this model is provided by the infrared spectra of metal carbonyl derivatives of the type $(\eta^6\text{-}C_6H_6)Cr(\eta^6\text{-}C_6F_5ML_n)$, where ML_n = $Re(CO)_5$ or $Fe(CO)_2(C_5H_5)$, prepared via lithiation of the pentafluorobenzene sandwich **61**.[53] Comparison with the corresponding uncomplexed $C_6F_5ML_n$ analogues revealed a marked reduction of the metal carbonyl stretching frequencies, ν_{CO}, in the sandwich compounds as the result of the enhanced back-donation to the metal carbonyl π^* manifold from the relatively electron-rich complexed pentafluorophenyl ring. In $(\eta^6\text{-}C_6H_6)Cr[(\eta^6\text{-}C_6F_5\text{-}Fe(CO)_2(C_5H_5)]$, **62**, the values of ν_{CO} are 2006 and 1963 cm^{-1}, whereas in C_6F_5–$Fe(CO)_2(C_5H_5)$ they are found at 2045 and 1997 cm^{-1}.[54]

Since that time, a number of other η^6-hexafluorobenzene-metal systems have been reported. By using an electron gun rather than resistive heating in a crucible, Timms was able to vaporise molybdenum and tungsten and co-condense them with hexafluorobenzene, either alone or in conjunction with other arenes. Thus, he was able to isolate $(\eta^6\text{-}C_6F_6)_2M$, and also $(\eta^6\text{-}C_6H_6)M(\eta^6\text{-}C_6F_6)$, where M = Mo, W;[55] bis(hexafluorobenzene)tungsten, **63**, yielded X-ray quality crystals but, as in the chromium case, the mixed benzene-hexafluorobenzene sandwich complexes suffered from disorder in the solid state. It was established that in $(\eta^6\text{-}C_6H_6)Cr(\eta^6\text{-}C_6F_6)$ the molecules stacked up the hexagonal axis of the unit cell in a head-to-tail fashion, but the inter- and intra-molecular separations of the benzene and hexafluorobenzene were essentially identical (~3.4 Å) resulting in an apparent positioning of a half-chromium atom between each pair of arene rings. However, upon lowering the symmetry by introducing a substituent, as in $(\eta^6\text{-}C_6H_6)Cr(\eta^6\text{-}C_6F_5PPh_2)$, this problem was resolved and the structure was secured.[56]

Co-condensation of hexafluorobenzene and mesitylene with either ruthenium or osmium atoms yielded the sandwich compound

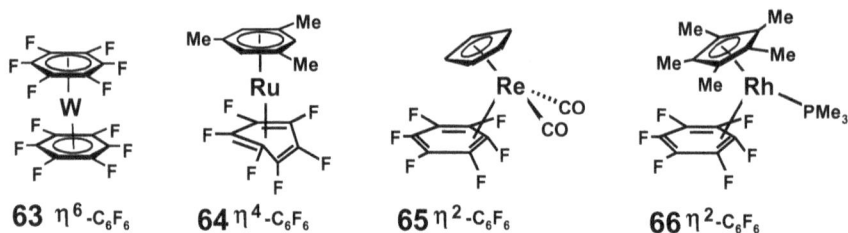

63 η^6-C_6F_6 **64** η^4-C_6F_6 **65** η^2-C_6F_6 **66** η^2-C_6F_6

Figure 8.12. Examples of η^6-, η^4- and η^2-bonded hexafluorobenzene metal complexes.

$(\eta^6$-$C_6H_3Me_3)M(\eta^4$-$C_6F_6)$, **64**,[57] in which the hexafluorobenzene ring has folded such that only four of the ring carbons are involved in coordination to the metal, thus satisfying the 18-electron rule. Using a different approach, Perutz found that photolysis of $(C_5H_5)Re(CO)_3$ in hexafluorobenzene resulted in loss of a carbonyl ligand and formation of $(C_5H_5)Re(CO)_2(\eta^2$-$C_6F_6)$, **65**, yet another coordination possibility for this fascinating arene (Figure 8.12). The structure of **65** was confirmed by X-ray crystallography, as was the related rhodium complex $(C_5Me_5)Rh(PMe_3)(\eta^2$-$C_6F_6)$, **66**.[58]

8.4.4. *Preparation of the Perfluorotropylium Cation*

Lemal has also reported the existence of the perfluorotropylium ion, $[C_7F_7]^+$, as a product of a series of sequential rearrangements beginning with perfluoronorbornadiene, **67**. As shown in Scheme 8.18, photolysis of **67** proceeded via the quadricyclane, **68**, and the bicyclo species **69**, leading eventually to perfluorocycloheptatriene, **70**, that exhibited a 2:2:2:2 ^{19}F NMR pattern. Finally, treatment with boron trifluoride etherate brought about abstraction of fluoride to form the salt $[C_7F_7]^+[BF_4]^-$; the singlet ^{19}F NMR signal for the perfluorotropylium cation, **71**, and its ready hydrolysis to form perfluorotropone, **72**, confirmed the assignment.[59]

8.4.5. *Pentafluorocyclopentadienyl-metal Complexes*

In contrast to the situation with hexafluorobenzene, for many years attempts to prepare metal complexes containing the

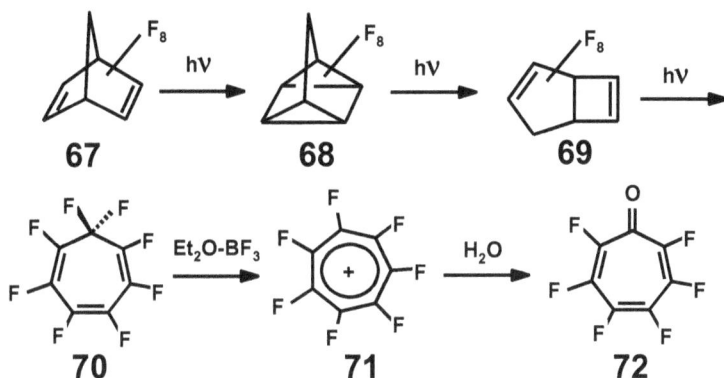

Scheme 8.18. Preparation and reactivity of the perfluorotropylium cation.

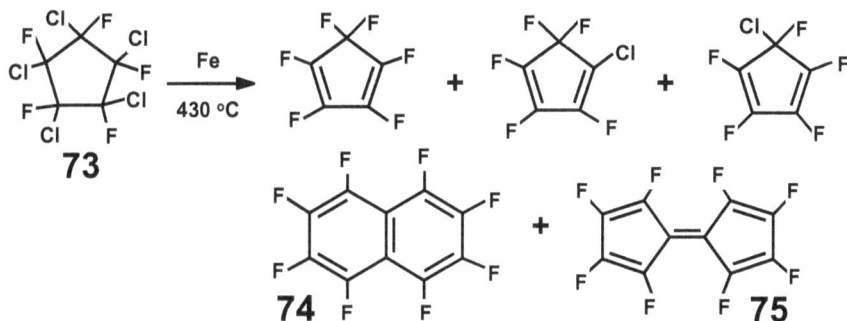

Scheme 8.19. Product mixture when $C_5Cl_5F_5$ is passed over iron at 430°C.

pentafluorocyclopentadienyl ligand were frustratingly unsuccessful. The synthesis of 1,2,3,4,5-pentafluorocyclopentadiene by zinc-mediated dechlorination of $C_5F_5Cl_5$, **73**, allowed the preparation of the thallium salt $Tl^+[C_5F_5]^-$ but attempts to form metal complexes containing Cr, Mo, W, Mn, Re, Fe or Ni were all unsuccessful.[60] However, one should also mention a much earlier report that had some relevance to this project.

When pentachloropentafluorocyclopentane, **73**, was pyrolysed by passage over mild steel at 430°C, among the products obtained after chromatographic separation were perfluorocyclopentadiene, a mixture of chlorofluorocyclopentenes and chlorofluorocyclopentadienes, perfluoronaphthalene, **74**, and traces of an intensely blue material (Scheme 8.19).[61] Knowing that an early (at that time unrecognised)

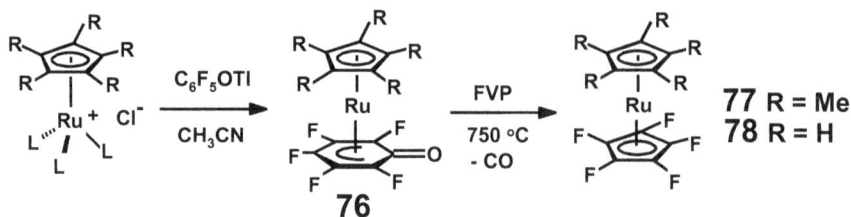

Scheme 8.20.   The first synthesis of a pentafluorocyclopentadienyl-metal complex.

synthesis of ferrocene involved passage of cyclopentadiene over heated iron,[62] there was much speculation that this highly coloured material might be perfluoroferrocene.[63] However, it was later reformulated as perfluorofulvalene, **75**, an isomer of perfluoronaphthalene; it is known that the corresponding perchloro- and perbromo-fulvalenes are also deep blue in colour.[64,65]

The initial breakthrough was finally achieved in 1992 by Russell Hughes at Dartmouth College. As shown in Scheme 8.20, the reaction of $[(C_5Me_5)Ru(CH_3CN)_3]Cl$ with the thallium salt of pentafluorophenol furnished the η^5-oxocyclohexadienyl derivative, **76**, whose ^{19}F NMR spectrum exhibited the required 2:2:1 intensity pattern. Flash vacuum pyrolysis (FVP) at 750°C resulted in loss of CO and formation of the first pentafluorocyclopentadienyl complex $(C_5Me_5)Ru(C_5F_5)$, **77**, which, gratifyingly, exhibited a single ^{19}F NMR resonance.[66]

Shortly thereafter, Hughes was able to prepare and obtain X-ray quality crystals of the parent sandwich compound, $(C_5H_5)Ru(C_5F_5)$, **78**; the structure revealed that the metal was situated closer to the fluorinated ring with ring-centroid-to-ruthenium distances of 1.716 Å and 1.847 Å, respectively.[67] This followed the pattern found previously in the chromium sandwich $(C_6H_6)Cr(C_6F_5PPh_2)$, in which the ring-centroid-to-chromium distances were 1.573 Å and 1.635 Å for the fluorinated and non-fluorinated rings, respectively.[56]

Despite these remarkable achievements, the decarbonylation approach was not successful in the ferrocene case, and another 20 years passed before this problem was resolved. Karlheinz Sünkel from Munich chose to use the technique of electrophilic fluorination,[68] whereby nucleophilic attack at fluorine with concomitant

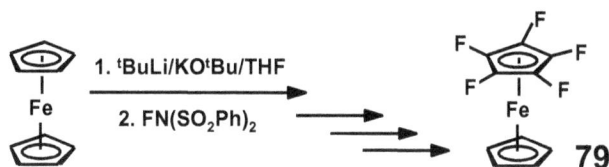

Scheme 8.21. Preparation of 1,2,3,4,5-pentafluoroferrocene.

displacement of an excellent leaving group is invoked. Perhaps the most widely used reagent industrially for electrophilic fluorinations is *SelectfluorTM*, developed by Eric Banks[69] at the University of Manchester, but in this case the reagent of choice was *N*-fluoro-*N*,*N*-bis(benzenesulfonate) imide.

Multiply repeated lithiation of ferrocene followed by reaction with $FN(SO_2Ph)_2$ led eventually to 1,2,3,4,5-pentafluoroferrocene, **79** (Scheme 8.21).[70] As in the earlier cases, the X-ray crystal structure revealed that the metal was positioned significantly closer to the fluorinated ring, but another factor was also discussed. As noted earlier, in $(C_6H_6)Cr(C_6F_6)$ the molecules stacked along a hexagonal axis in a head-to-tail fashion such that the inter- and intra-molecular separations of the benzene and hexafluorobenzene were essentially identical (~3.4 Å) resulting in an unresolvable disorder problem. In (C_5H_5) $Fe(C_5F_5)$, **79**, a similar stacking phenomenon occurs, but the intra-molecular distance between the C_5H_5 and C_5F_5 rings (3.26 Å) was slightly shorter than the intermolecular separation (3.37 Å), sufficiently different to allow resolution of the structure. So, although the bis-hexafluorobenzene derivatives of molybdenum and tungsten have been successfully obtained, bis(pentafluorocyclopentadienyl) sandwiches of iron, ruthenium or osmium remain as a challenge!

8.5. Concluding Remarks

Our objectives here have been twofold: first, to demonstrate the underlying significance of symmetry arguments in the elucidation of the mechanisms of chemical reactions and rearrangements; second, to show how the syntheses of molecules of unusually high symmetry continue to pose serious challenges. While the principles of the Conservation of Orbital Symmetry have provided a fundamental understanding of many aspects of chemical reactivity, experimental verification of these concepts has, in many cases, been achieved by the astute, sometimes rather subtle, breaking of molecular symmetry.

The early pioneers who added so much to our understanding of reaction mechanisms, and whose contributions are discussed in Chapter 2, frequently relied on their outstanding laboratory technical skills to acquire the necessary data, whereas today we depend heavily on the availability of the latest spectroscopic, spectrometric, and X-ray crystallographic instrumentation, together with powerful computational techniques. Nevertheless, we note that it is still easier to envisage a particular molecule, suitably labelled for our purposes, than it is to synthesise and characterise an analytically pure sample of it.

While many organic molecules exhibiting high or unusual symmetry are now available in kilogram quantities (cubanes, corannulene, etc.), others such as tri- or penta-prismane are still niche products, available only in small amounts, and with considerable difficulty. In contrast, in the inorganic domain where smaller bond angles are more easily tolerated, systems of tetrahedral, prismatic or cubic symmetry are legion.

Sometimes, however, Nature works in our favour. For example, adamantane possesses an idealised geometry with its 1.54 Å carbon-carbon bonds and 109.5° bond angles, and is the most energetically favoured isomer of $C_{10}H_{16}$, whereby almost any hydrocarbon of this formula, when treated with a strong Lewis acid, undergoes a multitude of carbocationic rearrangements so as to attain this structure (Figure 8.13).

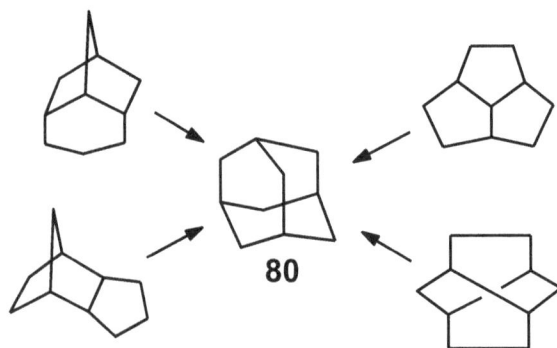

Figure 8.13. $C_{10}H_{16}$ isomers that rearrange to form adamantane, **80**.

Likewise, C_{60} is now available on a commercial scale merely by the vaporisation of graphite under appropriate conditions. Moreover, C_{60} has been found not only in ancient (**600** million years old) Russian geological samples, but also in interstellar space.

Let us conclude with the opening stanza of a poem published in 1794 by William Blake (1757–1827), the brilliant English artist, poet and mystic, a profound thinker and visionary.

> *Tyger, Tyger burning bright,*
> *In the forests of the night:*
> *What immortal hand or eye,*
> *Could frame thy fearful symmetry?*

Clearly, as chemists we should not fear symmetry but instead appreciate its awesome power, beauty and utility.

References

1. R. Chauvin, "Carbomers". I. A general concept of expanded molecules. *Tetrahedron Lett.* **1995**, *36*, 397–400.
2. C. Dallaire, M.A. Brook, A.D. Bain, C.S. Frampton and J.F. Britten, Tetrakis[(trimethylsilyl)ethynyl] Group 14 metal derivatives: an examination of the electronic interaction between two Group 14 metals connected by an acetylene wire. *Can. J. Chem.* **1993**, *71*, 1676–1683.
3. K.S. Feldman, C.K. Weinreb, W.J. Youngs and J.D. Bradshaw, Preparation and some subsequent transformations of tetraethynylmethane. *J. Am. Chem. Soc.* **1994**, *116*, 9019–9026.
4. U.H.F. Bunz, New carbon-rich organometallic architectures based on cyclobutadienecobalt and ferrocene modules. *J. Organomet. Chem.* **2003**, *683*, 269–287.
5. R. Suzuki, H. Tsukuda, N. Watanabe, Y. Kuwatani and I. Ueda, Synthesis, structure and properties of 3,9,15-tri- and 3,6,9,12,15,18-hexasubstituted dodecadehydro[18]annulenes ($C_{18}H_3R_3$ and $C_{18}R_6$) with D_{6h} symmetry. *Tetrahedron* **1998**, *54*, 2477–2496.
6. V. Maraval, L. Leroyer, A. Harano, C. Barthes, A. Saquet, C. Duhayon, T. Shinmyuzo and R. Chauvin, 1,4-Dialkynylbutatrienes: synthesis, stability, and perspectives in the chemistry of carbo-benzenes. *Chem. Eur. J.* **2011**, *17*, 5086–5100.
7. C. Cocq, N. Saffron-Merceron, Y. Coppel, C. Poidevin, V. Maraval and R. Chauvin, carbo-Naphthalene: a polycyclic carbo-benzenoid fragment of α-graphyne. *Angew. Chem. Int. Ed.* **2016**, *55*, 15133–15136.

8. L.T. Scott, G.J. DeCieco, J.L. Hyun and G.J. Reinhardt, Cyclynes. Part 4. Pericyclynes of the order [5], [6], [7], and [8]. Simple convergent syntheses and chemical reactions of the first homoconjugated cyclic polyacetylenes. *J. Am. Chem. Soc.* **1985**, *107*, 6546–6555.

9. C. Zou, C. Duhayon, V. Maraval and R. Chauvin, Hexasilylated total carbomer of benzene. *Angew. Chem. Int. Ed.* **2007**, *46*, 4337–4341.

10. H.A. Staab and F. Binnig, Synthesis and properties of hexa-*m*-phenylene. *Tetrahedron Lett.* **1964**, *5*, 319–321.

11. W. Zhang, S.M. Brombosz, J.L. Mendoza and J.S. Moore, A high yield, one-step synthesis of *o*-phenylene ethynylene cyclic trimer via precipitation-driven alkyne metathesis. *J. Org. Chem.* **2005**, *70*, 10198–10201.

12. V. Maraval and R. Chauvin, From macrocyclic oligo-acetylenes to aromatic ring *Carbo*-mers. *Chem. Rev.* **2006**, *106*, 5317–5343.

13. K. Cocq, C. Lepetit, V. Maraval and R. Chauvin, "*Carbo*-aromaticity" and novel *carbo*-aromatic compounds. *Chem. Soc. Rev.* **2015**, *44*, 6535–6559.

14. P. Manini, W. Manrein, V. Gramlich and F. Diederich, Expanded cubane: synthesis of a cage compound with a C_{56} core by acetylenic scaffolding and gas-phase transformations into fullerenes. *Angew. Chem. Int. Ed.* **2002**, *41*, 4339–4342.

15. I. Erden, A simple synthesis of [3]rotane. *Synth. Commun.* **1986**, *16*, 117–121.

16. R. Boese, T. Miebach and A. de Meijere, [3]Rotane: crystal structure, X-X difference electron density, and phase transition. *J. Am. Chem. Soc.* **1991**, *113*, 1743–1748.

17. J.M. Conia and J.M. Denis, Study of the rotanes (II). Tetracyclopropyliene (tetraspiro[2.0.2.0.2.0.2.0]dodecane. *Tetrahedron Lett.* **1969**, *10*, 3545–3546.

18. L. Fitjer, Coupling reactions of vinylidenecyclopropanes doubly halogenated in allylic positions – a productive synthesis of trispiro[2.0.2.0.2.0]nonane [3] rotane. *Angew. Chem. Int. Ed. Engl.* **1976**, *15*, 762–763.

19. L. Fitjer, A universal rotane synthesis – hexaspiro[2.0.2.0.2.0.2.0.2.0.2.0] octadecane([6]rotane). *Angew. Chem. Int. Ed. Engl.* **1976**, *15*, 763–764.

20. L. Fitjer and D. Wehle, Multiple cyclopropylmethyl-cyclobutyl rearrangements in a pentaspirohexadecanol. *Angew. Chem. Int. Ed. Engl.* **1979**, *18*, 868–869.

21. L. Fitjer and D. Wehle, Heptacyclo[19.3.0.0.01,5.05,9.09,13.013,17.017,21] tetracosane [6.5]coronane. *Angew. Chem. Int. Ed. Engl.* **1987**, *26*, 130–132.

22. L. Fitjer, C. Steeneck, S. Gaini-Rahimi, U. Schröder, K. Justus, P. Puder, M. Dittmer, C. Hassler, J. Weiser, M. Noltemeyer and M. Teichert, A new rotane family: synthesis, structure, conformation, and dynamics of [3.4]-, [4.4]-, [5.4]-, and [6.4]rotane. *J. Am. Chem. Soc.* **1998**, *120*, 317–328.

23. A. de Meijere, S.I. Kozhushkov, T, Haumann, R. Boese, C. Puls, M.J. Cooney and L.T. Scott, Completely spirocyclopropanated macrocyclic oligoacetylenes: the family of "exploding" [*n*]rotanes. *Chem. Eur. J.* **1995**, *1*, 124–131.

24. A. de Meijere and S.I. Kozhushkov, Completely spirocyclopropanated macrocyclic oligoacetylenes and their permethylated analogues: preparation and properties. *Chem. Eur. J.* **2002**, *8*, 3195–3202.

25. B. Busch and K. Dehnicke, The vibrational spectra of tetra-cyclopropyl compounds of silicon, germanium and tin. *J. Organomet. Chem.* **1974**, *67*, 237–242.

26. S.I. Kozhushkov, R.R. Kostikov, A.P. Molchanov, R. Boese, J. Benet-Buchholz, P.R. Schreiner, C. Rinderspacher, I. Ghiviringa and A. de Meijere, Tetracyclopropylmethane: a unique hydrocarbon with S_4 symmetry. *Angew. Chem. Int. Ed.* **2001**, *40*, 180–183.

27. L.C.F. Chao, A. Decken, J.F. Britten and M.J. McGlinchey, Organic and inorganic templates bearing $CCo_3(CO)_9$ cluster fragments: X-ray crystal structure of $C(CH_2OCOCCl_3)_4$ and of $CH_3C(O)CHC(OH)ML_n$, where $ML_n = CCo_3(CO)_9$ or $(C_6H_5)Cr(CO)_3$. *Can. J. Chem.* **1995**, *73*, 1106–1205.

28. T. Amaya and T. Hirao, Chemistry of sumanene. *Chem. Rec.* **2015**, *15*, 310–321.

29. S. Alvi and R. Ali, Synthetic approaches to bowl-shaped π-conjugated sumanene and its congeners. *Beilstein J. Org. Chem.* **2020**, *16*, 2212–2259.

30. Y.T. Yu and J.S. Siegel, Aromatic molecular-bowl hydrocarbons: synthetic derivatives, their structures and physical properties. *Chem. Rev.* **2006**, *106*, 4843–4867.

31. D. Meng, G. Liu, C. Xiao, Y. Shi, L. Zhang, L. Jiang, K.K. Baldridge, Y. Li, J.S. Siegel and Z. Wang, Corannurylene pentapetalae. *J. Am. Chem Soc.* **2019**, *141*, 5420–5406.

32. S. Brydges, L.E. Harrington and M.J. McGlinchey, Sterically hindered organometallics: multi-n-rotor (n = 5, 6 and 7) molecular propellers and the search for correlated rotations. *Coord. Chem. Res.* **2002**, *233*, 75–105, and references therein.

33. S. Brydges, J.F. Britten, L.C.F. Chao, H.K. Gupta, M.J. McGlinchey and D.L. Pole, The structure of a seven-bladed propeller: $C_7Ph_7^+$ is not planar. *Chem. Eur. J.* **1998**, *4*, 1201–1205.

34. M.A. Battiste, Heptaphenyltropilidene – product of a Diels-Alder reaction of triphenylcyclopropene. *Chem. Ind.* **1961**, 550–551.

35. M. Tamm, B. Dressel and R. Fröhlich, Molecular structure of a heptadentate cogwheel: $C_7Me_7^+$ is not planar. *J. Org. Chem.* **2000**, *65*, 6795–6797.

36. R.B. King and M.B. Bisnette, Organometallic chemistry of the transition metals XXI. Some π-pentamethylcyclopentadienyl derivatives of various transition metals. *J. Organomet. Chem.* **1967**, *8*, 287–297.

37. L.D. Field, T.W. Hambley, P.A. Humphrey, C.M. Lindall, G.J. Gainsford, A.F. Masters, T.G. StPierre and J. Webb, Decaphenylferrocene. *Aust. J. Chem.* **1995**, *48*, 851–860.

38. R.W. Heo and T.R. Lee, Ferrocenophanes with all carbon bridges. *J. Organomet. Chem.* **1999**, *578*, 31–42.

39. F.L. Hedberg and H. Rosenberg, Preparation and reactions of decachloroferrocene and decachlororuthenocene. *J. Am. Chem. Soc.* **1973**, *95*, 870–875.

40. C.H. Winter, Y.-H. Han, R.L. Ostrander and A.L. Rheingold, Decamercuration of ruthenocene. *Angew. Chem. Int. Ed. Engl.* **1993**, *32*, 1161–1163.

41. I.R. Butler, 1,2,3,4,5-Pentabromoferrocene and related compounds: a simple synthesis of useful precursors. *Inorg. Chem. Commun.* **2008**, *11*, 484–486.

42. K.J. Reimer and A. Shaver, pentachlorocyclopentadienyl derivatives of manganese and rhodium. *Inorg. Chem.* **1975**, *14*, 2707–2716.

43. M.J. Gerace, D.M. Lemal and H. Ertl, Tetrafluorocyclobutadiene. *J. Am. Chem. Soc.* **1975**, *97*, 5584–5586.

44. D.J. Cook, M. Green, N. Mayne and F.G.A. Stone, Chemistry of metal carbonyls. 48. Complexes of Dewar hexafluorobenzene. *J. Chem. Soc. A* **1968**, 1771–1775.

45. P.L. Timms, Formation of complexes from transition-metal vapours. *Chem. Commun.* **1968**, 1525.

46. P.S. Skell and M.J. McGlinchey, Reactions of transition metal atoms with organic substrates. *Angew. Chem. Int. Ed. Engl.* **1975**, *14*, 195-199.

47. M.J. McGlinchey, Synthesis of organometallic compounds using metal atoms, in *The Chemistry of the Metal-Carbon Bond*; Wiley-VCH: New York, USA, 1982; pp. 539-574.

48. P.S. Skell, D.L. Williams-Smith and M.J. McGlinchey, Chromium atoms in organometallic synthesis. *J. Am. Chem. Soc.* **1973**, *95*, 3337-3340.

49. E.O. Fischer and W. Hafner, Di-benzene-chromium – concerning aromatic complexes of metals. *Z. Naturforsch.* **1955**, *10b*, 665–668.

50. R. Middleton, J.R. Hull, S.R. Simpson, C.H. Tomlinson and P.L. Timms, Chemistry of transition-metal vapors. 3. Formation of complexes with arenes, trifluorophosphine and nitric oxide. *J. Chem. Soc. Dalton Trans.* **1973**, 120–124.

51. M.J. McGlinchey and T-S. Tan, Nucleophilic substitution reactions in fluorinated bis(arene)chromium complexes. *J. Am. Chem. Soc.* **1976**, *98*, 2271-2275.

52. D. Seyferth, Bis(benzene)chromium. 2. Its discovery by E.O. Fischer and W. Hafner and subsequent work by the research groups of E.O. Fischer, H.H. Zeiss, C. Elschenbroich, and others. *Organometallics* **2002**, *21*, 2800–2820.

53. A. Agarwal, M.J. McGlinchey and T-S. Tan, The synthesis and chemistry of 1,2,3,4,5-pentafluorochromarene: Electronic effect of a π-Cr(C$_6$H$_6$) moiety. *J. Organomet. Chem.* **1977**, *141*, 85-97.

54. R.B. King and M.B. Bisnette, Reactions of alkali metal derivatives of metal carbonyls. 5. Perfluoroaryl derivatives of iron prepared by reaction between

NaFe(CO)$_2$(C$_5$H$_5$) and certain aromatic fluorocarbons. *J. Organomet. Chem.* **1964**, *2*, 38–43.

55. J.J. Barker, A.G. Orpen, A.J. Seeley and P.L. Timms, Crystal structure of [W(η-C$_6$F$_6$)$_2$] synthesised together with other sandwich complexes of tungsten and molybdenum containing η^6-C$_6$F$_6$ or η^6-C$_6$H$_3$F$_3$–1,3,5 ligands from atoms of the metals. *J. Chem. Soc. Dalton Trans.* **1993**, 3097–3102.

56. R. Faggiani, N. Hao, C.J.L. Lock, B.G. Sayer and M.J. McGlinchey, Unsymmetrical sandwich compounds: The preparation and characterisation of (η^6-benzene)(η^6-1,2,3,4,5-pentafluoro-6-diphenylphosphinobenzene)-chromium and its reaction with [Rh(CO)$_2$Cl]$_2$. *Organometallics* **1983**, *2*, 96-100.

57. A. Martin, A.G. Orpen, A.J. Seeley and P.L. Timms, Synthesis of new η^4-hexafluorobenzene complexes of ruthenium and osmium from atoms of the metals: crystal structure of [Ru(η^6-C$_6$H$_3$Me$_3$–1,3,5)(η^4-C$_6$F$_6$)]. *J. Chem. Soc. Dalton Trans.* **1994**, 2251–2255.

58. C.L. Higgitt, A.H. Klahn, M.H. Moore, B. Oelekers, M.G. Partridge and R.N. Perutz, Structure and dynamics of the η^2-hexafluoro benzene complexes [Re(η^5-C$_5$H$_4$R)(CO)$_2$(η^2-C$_6$F$_6$) (R = H or Me) and [Rh(η^5-C$_5$Me$_5$)(PMe$_3$)(η^2-C$_6$F$_6$). *J. Chem. Soc. Dalton Trans.* **1997**, 1269–1280.

59. W.P. Dailey and D.M. Lemal, Perfluorotropilidene valence isomers and the per-fluorotropylium ion. *J. Am. Chem Soc.* **1984**, *106*, 1169–1170.

60. G. Paprott, S. Lehmann and K. Seppelt, Reactions of 1,2,3,4,5-pentafluorocyclo-pentadiene. *Chem. Ber.* **1988**, *121*, 727–733.

61. R.E. Banks, M. Bridge, R.N. Haszeldine, D.W. Roberts and N.I. Tucker, Polyfluorocyclopentadienes. Part VI. Synthesis of 1- and 5-chloropentafluorocyclo-pentadiene. *J. Chem. Soc.* **1970**, 2531–2535.

62. S.A. Miller, J.A. Tebboth and J.F. Tremaine, Dicyclopentadienyliron. *J. Chem. Soc.* **1952**, 632–635.

63. The present author, who was at that time a beginning graduate student working on a different project in the same laboratory, was a spectator during this experiment. He well remembers the excitement and much speculation among the experimenters as to whether the intensely blue compound was perfluoroferrocene.

64. V. Mark, Perchlorofulvalene. *Tetrahedron Lett.* **1961**, *2*, 333–336.

65. R. West and P.T. Kwitowski, Reactions of hexabromocyclopentadiene and the synthesis of octabromofulvalene. *J. Am. Chem. Soc.* **1968**, *90*, 4697–4701.

66. O.J. Curnow and R.P. Hughes, [Ru(η^5-C$_5$Me$_5$)(η^5-C$_5$F$_5$)]: the first transition metal complex containing a perfluorocyclopentadienyl ligand. *J. Am. Chem. Soc.* **1992**, *114*, 5895–5898.

67. R.P. Hughes, X. Zheng, R.L. Ostrander and A.L. Rheingold, Synthesis and molecular structure of [Ru(η^5-C$_5$H$_5$)(η^5-C$_5$F$_5$)]. Intramolecular structural comparison of the cyclopentadienyl ligand and its perfluorinated analogue. *Organometallics* **1994**, *13*, 1567–1568.

68. T. Umemoto, Y. Yang and G.B. Hammond, Development of N-F fluorinating agents and their fluorinations: historical perspective. *Beilstein. J. Org. Chem.* **2021**, *17*, 1752–1813.

69. R.E. Banks, SelectfluorTM reagent-TEDA-BF4 in action: tamed fluorine at your service. *J. Fluor. Chem.* **1998**, *87*, 1–17.

70. K. Sünkel, S. Weigand, A. Hoffmann, S. Blomeyer, C.G. Reuter, V. Vishnevsky and N.W. Mitzel, Synthesis and characterisation of 1,2,3,4,5-pentafluoroferro-cene. *J. Am. Chem. Soc.* **2015**, *137*, 126–129.

Appendix

A Brief Introduction
to Symmetry Point Groups

"An amino acid residue (other than glycine) has no symmetry elements. The general operation of conversion of one residue of a single chain into a second residue equivalent to the first is accordingly a rotation about an axis accompanied by translation along the axis. Hence the only configurations for a chain compatible with our postulate of equivalence of the residues are helical configurations."

— *Linus Pauling*

At the outset, one should emphasise that this additional section is merely a brief summary (or reminder) of the use of point group nomenclature as a shorthand to categorise the symmetry properties of an object, in this case always a molecule. We would certainly encourage any reader wishing to delve more deeply into the topic to peruse one of the many excellent textbooks on the chemical applications of Group Theory and symmetry in such areas as vibrational spectroscopy, molecular orbital calculations, ligand field theory, X-ray crystallography, etc.

We can classify molecular structures in terms of their *symmetry elements* — mirror planes (σ), centre of symmetry (i), proper (C_n) and improper (S_n) axes — and their associated *symmetry operations* — reflection, inversion, rotation, and rotation-reflection, respectively.

Reflection in a mirror and inversion about a centre (in a Cartesian system, the x, y, z coordinates for each atom become $-x$, $-y$, $-z$) are relatively self-evident. However, for proper rotations one should indicate how many times this operation must be carried out to return to the original situation; thus, each rotation about a C_3 axis moves the relevant atoms through 120°. The operation about an improper axis can be thought of as a two-step process, whereby an initial proper rotation is then followed by reflection through a plane perpendicular to the axis; a number of examples will illustrate this procedure.

To determine the point group of any given molecule, one can proceed via a series of simple questions that eventually lead to a definitive answer. Let us begin with several easy situations. We can all immediately recognise tetrahedral, octahedral and icosahedral structures, as represented by tetrabromomethane or white phosphorus (P_4), sulfur hexafluoride or C_8H_8 (cubane), and $C_{20}H_{20}$ (dodecahedrane) or the *closo*-dodecahydrododecaborate anion $[B_{12}H_{12}]^{2-}$ (Figure A.1); these are designated as T_d, O_h and I_h, respectively.

We now consider systems possessing a proper axis, C_n, and ask whether there are other C_2 axes perpendicular to the principal axis (the one of highest order, normally assigned as the z axis). When the answer is negative, we next look for mirror planes; if there is one in the xy plane, i.e., orthogonal to the z axis, it is designated as a

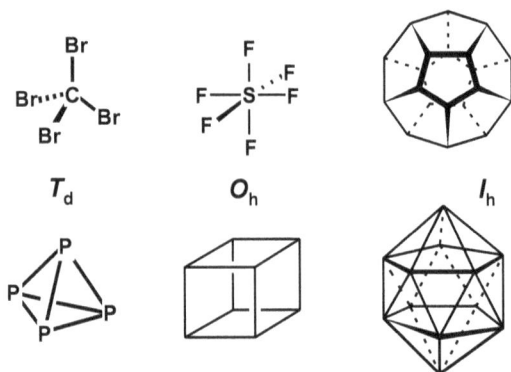

Figure A.1. A selection of highly symmetrical molecules: CBr_4, P_4, SF_6, C_8H_8 (cubane), $C_{20}H_{20}$ (dodecahedrane) and $[B_{12}H_{12}]^{2-}$.

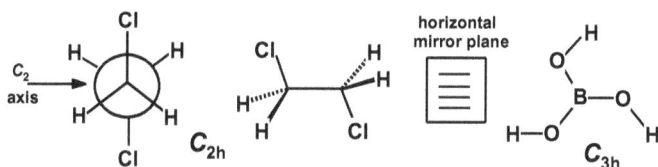

Figure A.2. Selected molecules possessing C_{nh} symmetry.

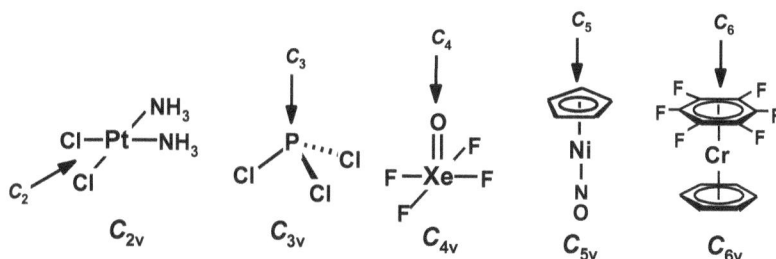

Figure A.3. A selection of molecules possessing C_{nv} symmetry.

horizontal plane (σ_h), and the molecule has C_{nh} symmetry. Typical examples are staggered 1,2-dichloroethane, C_{2h}, or boric acid $B(OH)_3$ in a planar conformation, C_{3h} (Figure A.2).

If, instead, there are only mirror planes that contain the axis, these are designated as vertical planes (σ_v), and are represented by such molecules as $PtCl_2(NH_3)_2$ (*cis*-platin, the widely used anticancer drug) with its two-fold axis bisecting the Cl–Pt–Cl angle, the pyramidal structure of phosphorus trichloride, xenon(VI) oxyfluoride $F_4Xe=O$, which has a C_4 axis containing the xenon-oxygen bond, (η^5-cyclopentadienyl)nickelnitrosyl $(C_5H_5)NiNO$, and (η^6-benzene)(η^6-hexafluorobenzene)chromium(0) $(C_6F_6)Cr(C_6H_6)$, in which the C_5 and C_6 axes, respectively, pass through the midpoints of the rings and also the metal atom. Their point groups are C_{2v}, C_{3v}, C_{4v}, C_{5v}, and C_{6v}, respectively (Figure A.3).

Nevertheless, one must take care in certain cases. For example, as depicted in Figure A.4, 1,3,5,7,9,11-hexabromocoronene is planar and exhibits C_{6h} symmetry. However, the apparently closely analogous 1,3,5,7,9-pentabromocorannulene is bowl-shaped, lacks any

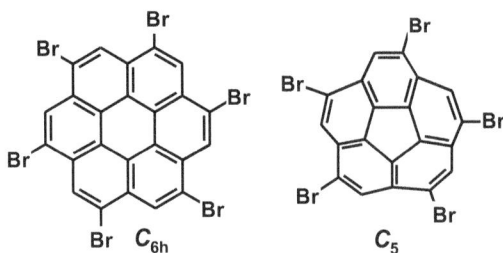

Figure A.4. Comparison of polycyclic molecules possessing five-fold and six-fold symmetry.

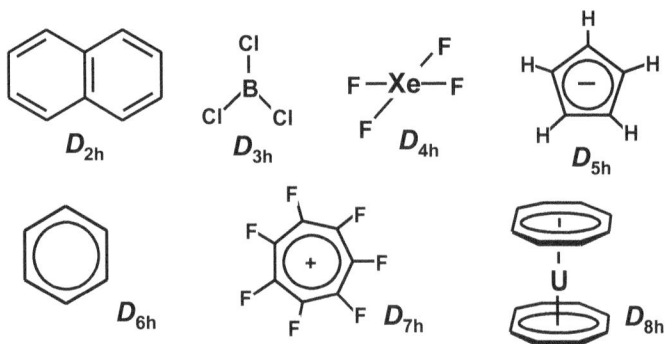

Figure A.5. Molecules possessing D_{nh} symmetry.

mirror planes in its ground state structure, and is chiral; its point group is therefore C_5 and the molecule only attains C_{5h} symmetry when bowl-to-bowl inversion passes through a planar transition state.

Returning now to molecules possessing nC_2 axes perpendicular to the C_n principal axis, we again ask about the existence of mirror planes. When the molecule has a horizontal plane (σ_h), as in naphthalene, boron trichloride, xenon tetrafluoride, the cyclopentadienide anion, benzene, the perfluorotropylium cation $[C_7F_7]^+$, and eclipsed uranocene $(C_8H_8)_2U$, the point groups are D_{2h}, D_{3h}, D_{4h}, D_{5h}, D_{6h}, D_{7h} and D_{8h}, respectively (Figure A.5).

However, in cases where there is no σ_h but instead only vertical planes that lie *between* the C_2 axes, the molecule is now assigned the point group D_{nd}. This is illustrated in Figure A.6 for staggered

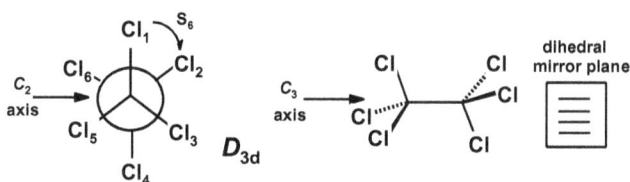

Figure A.6. The orientation of the C_3 and C_2 axes and of a dihedral mirror plane in D_{3d}-symmetric staggered C_2Cl_6.

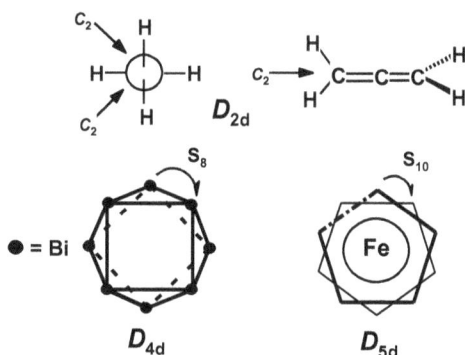

Figure A.7. A selection of molecules possessing D_{nd} symmetry.

hexachloroethane (D_{3d}) in which the three so-called *dihedral planes* (σ_d) each contain two chlorines and both carbons. Such molecules now possess improper axes (S_n) whereby one can visualise rotation of an atom through $360°/n$ and then reflect into the other half of the molecule. To make it absolutely clear, let us begin with $Cl(1)$: rotate clockwise through $60°$ and then reflect to become $Cl(2)$; this sequence of S_6 operations repeats until we return to $Cl(1)$.

Other examples, shown in Figure A.7, are allene $H_2C=C=CH_2$, with its orthogonally oriented termini, the square antiprismatic octabismuth cation $[Bi_8]^{2+}$ (D_{4d}), with its S_8 improper axis, and also staggered ferrocene that possesses an S_{10} improper axis, and whose point group is D_{5d}.

When the system possesses a principal C_n axis and nC_2 axes but no mirror planes, the symmetry is lowered to D_n, as in Werner's famously

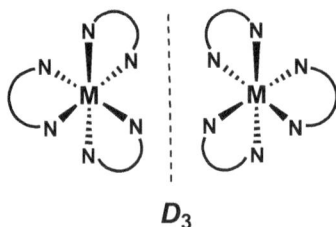

D_3

Figure A.8. Enantiomers of the $[Co(ethylenediamine)_3]^{3+}$ cation.

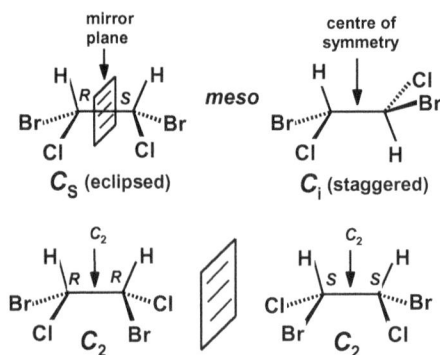

Figure A.9. Eclipsed/staggered rotamers of *meso* 1,2-dibromo-1,2-dichloroethane, and the enantiomers of the C_2-symmetric isomer.

chiral D_3-symmetric $[Co(en)_3]^{3+}$ cation that he prepared as part of his demonstration of the octahedral nature of hexacoordinate metal complexes (Figure A.8).

The various isomers and conformers of 1,2-dibromo-1,2-dichloroethane yield a number of different structures that are readily identified by the assignment of point groups. Beginning with the *meso* isomer, the fully eclipsed rotamer shown in Figure A.9 has no rotational axes, but does possess a molecular mirror plane; this is designated C_S. (Other examples of C_S-symmetric molecules include $Cl_2S=O$, $PClF_2$ and HOD.) In contrast, its fully staggered counterpart has only a centre of symmetry and is assigned the point group C_i (Figure A.9). The *d,l* isomers of CHBrCl–CHBrCl are also shown in Figure A.9 which reveals that they possess only a C_2 axis, and are thus chiral. It is noteworthy that

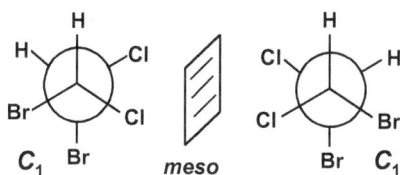

Figure A.10. Chiral rotamers of *meso* 1,2-dibromo-1,2-dichloroethane that have C_1 symmetry.

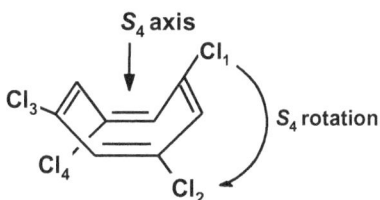

Figure A.11. 1,3,5,7-tetrachlorocyclooctatetrene has no centre of symmetry or mirror plane but does possess a four-fold improper axis and is achiral.

the assignment of the point group in the *meso* isomer is dependent on the identity of the rotamer, and indeed it loses all its symmetry elements during the transition from eclipsed to staggered conformations; this is designated C_1 (Figure A.10). In contrast, in the chiral (*d,l*) version its C_2 character is always conserved, no matter the conformation.

Indeed, one can simplify the widely cited criteria for chirality, i.e., the non-superimposability of its mirror image brought about through lack of a mirror plane or centre of symmetry. A little thought reveals that an S_1 operation (rotation through 360° followed by reflection) has the same result as a simple reflection. Moreover, an S_2 operation (rotation through 180° followed by reflection) is identical to inversion through a centre of symmetry. Hence, a more straightforward criterion for chirality is lack of an improper axis. This is particularly evident in, for example, 1,3,5,7-tetrachlorocyclooctatetraene, as depicted in Figure A.11. This molecule has neither a mirror plane nor an inversion centre, but it does possess an S_4 axis; it is achiral and superimposable on its mirror image. (Build a model and see for yourself.)

Finally, in linear molecules, which necessarily possess a C_∞ axis, we distinguish two situations: in dinitrogen N≡N, beryllium dichloride Cl–Be–Cl, or in carbon suboxide O=C=C=C=O, there is a centre of symmetry and a mirror plane orthogonal to the axis; this is designated $D_{\infty h}$. In contrast, in hydrogen cyanide H–C≡N, or chloroethyne H–C≡C–Cl, there is no inversion centre or horizontal plane, but an infinite number of vertical mirror planes that contain the axis, leading to the label $C_{\infty v}$.

To summarise:

(1) Look for special cases: linear, tetrahedral, octahedral or icosahedral.
(2) Is there a single C_n axis? Does it have a horizontal plane? Yes, then it is C_{nh}. Does it instead only have vertical planes? Yes, then it is C_{nv}. If it has no mirror planes, then it is C_n (chiral).
(3) Are there multiple C_2 axes perpendicular to the principal C_n axis? Does it have a horizontal plane? Yes, then it is D_{nh}. Does it instead only have vertical (dihedral) planes? Yes, then it is D_{nd}. If it has no mirror planes, then it is D_n (chiral).
(4) If there are no proper rotation axes, is there a single mirror plane? Yes, then it is C_s. Is there only a centre of symmetry? Yes, then it is C_i. If there are no elements of symmetry, then it is C_1 (chiral).

This will cover most of the commonly encountered point groups, but remember to look for whether there remains an improper (rotation-reflection) axis, then S_n (achiral).

Index

271

www.ingramcontent.com/pod-product-compliance
Lightning Source LLC
Chambersburg PA
CBHW050546190326
41458CB00007B/1937